GRASPING THE MOMENT

Sensemaking in Response to Routine Incidents and Major Emergencies

GRASPING THE MOMENT

Sensemaking in Response to Routine Incidents and Major Emergencies

Christopher Baber
Richard McMaster

CRC Press
Taylor & Francis Group
Boca Raton London New York

CRC Press is an imprint of the
Taylor & Francis Group, an **informa** business

CRC Press
Taylor & Francis Group
6000 Broken Sound Parkway NW, Suite 300
Boca Raton, FL 33487-2742

© 2016 by Taylor & Francis Group, LLC
CRC Press is an imprint of Taylor & Francis Group, an Informa business

No claim to original U.S. Government works

Printed on acid-free paper
Version Date: 20160701

International Standard Book Number-13: 978-1-4724-7080-5 (Hardback)

This book contains information obtained from authentic and highly regarded sources. Reasonable efforts have been made to publish reliable data and information, but the author and publisher cannot assume responsibility for the validity of all materials or the consequences of their use. The authors and publishers have attempted to trace the copyright holders of all material reproduced in this publication and apologize to copyright holders if permission to publish in this form has not been obtained. If any copyright material has not been acknowledged please write and let us know so we may rectify in any future reprint.

Except as permitted under U.S. Copyright Law, no part of this book may be reprinted, reproduced, transmitted, or utilized in any form by any electronic, mechanical, or other means, now known or hereafter invented, including photocopying, microfilming, and recording, or in any information storage or retrieval system, without written permission from the publishers.

For permission to photocopy or use material electronically from this work, please access www.copyright.com (http://www.copyright.com/) or contact the Copyright Clearance Center, Inc. (CCC), 222 Rosewood Drive, Danvers, MA 01923, 978-750-8400. CCC is a not-for-profit organization that provides licenses and registration for a variety of users. For organizations that have been granted a photocopy license by the CCC, a separate system of payment has been arranged.

Trademark Notice: Product or corporate names may be trademarks or registered trademarks, and are used only for identification and explanation without intent to infringe.

Library of Congress Cataloging-in-Publication Data

Names: Baber, Christopher, 1964- author. | McMaster, Richard Benjamin, author.
Title: Grasping the moment : sensemaking in response to "routine" emergencies and major incidents / by Chris Baber and Richard McMaster.
Description: Burlington, VT : Ashgate, [2016] | Includes bibliographical references and index.
Identifiers: LCCN 2015046641 (print) | LCCN 2016019514 (ebook) | ISBN 9781472470805 (hardback : alk. paper) | ISBN 9781472470812 (ebook) | ISBN 9781472470829 (epub)
Subjects: LCSH: Emergency management--Decision making. | Crisis management. | Organizational behavior. | Law enforcement. | Knowledge management.
Classification: LCC HD49 .B33 2016 (print) | LCC HD49 (ebook) | DDC 658.4/056--dc23
LC record available at https://lccn.loc.gov/2015046641

Visit the Taylor & Francis Web site at
http://www.taylorandfrancis.com

and the CRC Press Web site at
http://www.crcpress.com

Printed and bound in the United States of America by Publishers Graphics, LLC on sustainably sourced paper.

For Ted Megaw who always made unusual sense.

CB

For E, C and R who are wonderfully nonsensical.

RM

Contents

LIST OF FIGURES		xiii
LIST OF TABLES		xvii
PREFACE		xix
ABOUT THE AUTHORS		xxi

CHAPTER 1 INTRODUCTION — 1
- 1.1 The Challenges of Sensemaking — 1
  - 1.1.1 When Does Sensemaking Happen? — 5
  - 1.1.2 What Is Made in Sensemaking? — 6
- 1.2 Common Ground in Conversations — 8
  - 1.2.1 Sensemaking and Common Ground — 11
- 1.3 Three Types of Sensemaking — 12
  - 1.3.1 Individual Sensemaking — 13
  - 1.3.2 Artefact-Driven Sensemaking — 13
  - 1.3.3 Collaborative Sensemaking — 14
- 1.4 Macrocognition and Sensemaking — 15
- 1.5 Distributed Cognition as a Unifying Concept for Sensemaking — 16
- 1.6 Graphically Representing Distributed Cognition — 17
- 1.7 A Note on *Emergency Response* — 20

CHAPTER 2 INDIVIDUAL SENSEMAKING — 21
- 2.1 Introduction — 21
- 2.2 Representing Prior Experience — 23
  - 2.2.1 RPD Making — 24
- 2.3 Seeing the Gaps: The Data-Frame Model — 27

VII

	2.4	Flexecution	30
	2.5	The Problem of Bias	31
		2.5.1 Frames and Biases	32
		2.5.2 Sources of Bias in Sensemaking	34
	2.6	Conclusions	36
Chapter 3		**Sensemaking with Artefacts**	39
	3.1	Introduction	39
	3.2	Artefacts as External Representations	42
	3.3	Artefacts as Part of a Cognitive System	43
	3.4	Artefacts as Resources for Action	45
	3.5	The Problem with Sensemaking as *Representation Construction*	49
	3.6	Distributed Cognition and the Extended Mind	52
		3.6.1 The Mark of the Cognitive	53
	3.7	Conclusions	55
Chapter 4		**Collaborative Sensemaking**	57
	4.1	Introduction	57
	4.2	Collaborative Search after Meaning	59
		4.2.1 Identity: Perception of the Environment Is Affected by the Perception of Self or Group	59
		4.2.2 Retrospective: Sensemaking Is Concerned with Making Sense of Events That Have Already Happened	60
		4.2.3 Enactment: The Process of Making Sense Necessitates Active Involvement with the Environment and the Situation	60
		4.2.4 Social: Making Sense Involves the Creation of Shared Meaning and Shared Experience That Guides Organizational Decision Making	61
		4.2.5 Ongoing: Sensemaking Is a Continuous Process That Starts before and Continues after an Event	61
		4.2.6 Extracted Cues: Information Is Provided by Interactions with the Environment; This Prompts Further Data Collection	62
		4.2.7 Plausible Rather than True: Sensemaking Generates a Coherent, Reasonable and Memorable Understanding of an Event That Guides Action, Rather than Attempting Accuracy	62
	4.3	The Problem of Situation Awareness	64
		4.3.1 Distributed SA	65
		4.3.2 The Role of Artefacts in Distributed SA	69
	4.4	Sensemaking as System Activity	70
	4.5	Conclusions	72

Contents

Chapter 5 Command and Control in the UK Emergency Services — 73
  5.1 Introduction — 73
  5.2 Emergency Service Operations in the United Kingdom — 74
  5.3 The Concept of C2 — 78
    5.3.1 Police Incident Response C2 Organisation — 79
    5.3.2 Fire and Rescue: The Incident Command Model — 82
    5.3.3 OODA Loop — 82
    5.3.4 A Generic Process Model of C2 — 84
  5.4 The Future of C2 — 86

Chapter 6 Sensemaking in Command and Control — 89
  6.1 Introduction — 89
  6.2 Collaborative Networks — 91
  6.3 Planning and Adaptation (Replanning) — 93
  6.4 Problem Detection — 96
  6.5 Co-Ordination — 98
  6.6 Conclusions — 99

Chapter 7 Managing Routine Incidents — 101
  7.1 Introduction — 101
  7.2 Making Sense of the Call — 104
  7.3 Supporting Responding Units — 106
  7.4 Officer Attending — 114
  7.5 Closing the Incident — 120

Chapter 8 Distributed Cognition in Routine Incidents — 121
  8.1 Introduction — 121
  8.2 Allocating Resources to Incidents — 123
  8.3 Making Sense with Artefacts — 127
  8.4 Making Sense through Artefacts — 129
  8.5 Improvised Artefacts — 131
  8.6 Making Sense through Collaboration — 132
  8.7 Conclusions — 134

Chapter 9 Responding to Major Incidents — 137
  9.1 Introduction — 137
  9.2 Case Study: Walham Floods, 2007 — 141
    9.2.1 Background: The Defence of Walham Electricity Substation — 142
  9.3 C2 Structures — 143
  9.4 Keeping Track of the Changing Situation — 145
  9.5 Conclusions — 149

Chapter 10 Distributed Cognition in Major Incidents — 151
  10.1 Introduction — 151
  10.2 Common Ground — 155
  10.3 Making Sense with Artefacts — 160
  10.4 Making Sense through Artefacts — 161

	10.5	Making Sense through Collaboration	162
		10.5.1 Organisational Structures	162
		10.5.2 Social Processes and Collaborative Sensemaking	163
	10.6	Conclusions	165

CHAPTER 11 THE CHALLENGES OF INTEROPERABILITY — 167

11.1	Introduction	167
11.2	Defining Interoperability	168
	11.2.1 Technology	169
	11.2.2 SOPs	170
	11.2.3 Training	172
	11.2.4 Usage	173
11.3	Case Study: Initial Response to Terrorist Attacks on July 7, 2005	173
11.4	Case Study: Initial Response to Boston Marathon Bombings (2013)	176
11.5	Conclusions	178

CHAPTER 12 SENSEMAKING AND ORGANISATIONAL STRUCTURE IN EMERGENCY RESPONSE — 181

12.1	Introduction	181
12.2	Sensemaking as a Social Process	182
12.3	Analysing Network Structures and Interoperability	184
12.4	Conclusions	189
12.5	The Challenge of Sharing Information	190
	12.5.1 Data and Information Semantics	191
	12.5.2 Data Quality, Quantity and Timeliness	191
	12.5.3 Data and Information Relevance	192

CHAPTER 13 COMMON OPERATING PICTURES — 193

13.1	Introduction	193
13.2	Informing versus Understanding	194
13.3	Ontologies for COP	197
13.4	COP as a Representation of the State of the World or as a Collaborative Planning Tool?	203
13.5	Situation Space versus Decision Space	206
13.6	COP as Product versus COP as Process	206
	13.6.1 Routine Emergencies	208
13.7	Conclusions	210

CHAPTER 14 DISCUSSION — 211

14.1	Introduction	211
	14.1.1 Sensemaking during Routine Emergencies	212
	14.1.2 Breakdown of Sensemaking during a Major Incident	213
14.2	Reflections on Sensemaking Theories	214
	14.2.1 Individual Sensemaking	215

		14.2.2	Artefact-Driven Sensemaking	216
		14.2.3	Collaborative Sensemaking	217
		14.2.4	Sensemaking as Distributed Cognition	219
	14.3	Revisiting C2		221
	14.4	Future Directions for Incident Response		222
		14.4.1	Major Incidents	223
		14.4.2	Emergency Services Interoperability	225
	14.5	Final Words		225

REFERENCES 227

APPENDIX A: TIMELINES OF EVENTS IN LONDON BOMBINGS (2005) 241

APPENDIX B: SOCIAL NETWORK ANALYSIS CENTRALITY SCORES 245

INDEX 247

List of Figures

Figure 1.1	Task–artefact cycle	12
Figure 1.2	Macrocognitive framework	15
Figure 1.3	Graphically representing distributed cognition in the target review process	19
Figure 2.1	Strands of expert activity in the field	22
Figure 2.2	RPD model	25
Figure 2.3	Sensemaking in the data-frame model	29
Figure 3.1	Speed bugs on an airspeed indicator	45
Figure 3.2	Notes made on a police motorbike fuel tank during a traffic operation	48
Figure 3.3	Notional model of the sensemaking loop	50
Figure 4.1	Representing shared (overlapping) SA	67
Figure 4.2	Representing distributed SA	68
Figure 4.3	Dynamic systems model of intelligence analysis processes	70
Figure 5.1	Emergency services major incident command structure	74
Figure 5.2	From strategic intent to operational decision making	78
Figure 5.3	Schematic of emergency response in a regional control in a UK police force	80

LIST OF FIGURES

Figure 5.4	NDM	81
Figure 5.5	The fire and rescue incident command model	82
Figure 5.6	The Boyd Cycle/OODA loop in full detail	83
Figure 5.7	Generic process model of C2	85
Figure 7.1	Routine incident response tasks, personnel and lines of communication	103
Figure 7.2	A call handler's workstation	105
Figure 7.3	Annotated process flow for *taking a 999 call*	107
Figure 7.4	Controllers working together at adjacent workstations	108
Figure 7.5	Annotated process flow for *support responding units*	111
Figure 7.6	The sensemaking process involved in establishing the identity of a suspect	117
Figure 8.1	Representation of the call handler's sensemaking process	122
Figure 8.2	Controller's workstation	124
Figure 8.3	Screenshot of controller's open incident list	124
Figure 9.1	Floodwaters threaten the Walham electricity substation	142
Figure 9.2	Bronze commander's view of the adapted fire and rescue C2 structure	144
Figure 9.3	Interior of a fire and rescue ICU	145
Figure 9.4	An annotated fire and rescue entry control board	147
Figure 11.1	Interoperability Continuum	169
Figure 11.2	Cross-agency huddle at incident	171
Figure 11.3	Map showing the locations of the July 7 explosions	174
Figure 11.4	Time since explosion for emergency services to declare a major incident at the scenes, based on timings in the Report of the 7 July Review Committee	175
Figure 11.5	Location of explosions during Boston Marathon	177
Figure 12.1	Social network for the initial response to all incidents on 7/7	185
Figure 12.2	Sequence diagram showing the activity of different agencies on 7/7	187
Figure 13.1	Emergency services map system	196
Figure 13.2	Mind-map representation of CHALET acronym	198

Figure 13.3	Class diagram of information in an incident	199
Figure 13.4	Examples of user interfaces designed to support Figure 13.3	201
Figure 13.5	Incident log list	204
Figure 13.6	*Nesting* of talkgroups through use of scanning and talkgroup prioritisation	209
Figure 13.7	Data terminal in a Helsinki Police patrol vehicle (2011)	209

List of Tables

Table 1.1	Distributed cognition features of each activity within a task	18
Table 6.1	Tightly and loosely coupled work systems	91
Table 6.2	Potential impacts of different command arrangements on macrocognitive processes	92
Table 6.3	The characteristics of communities of practice and exploration networks	93
Table 8.1	Key incident response tasks, personnel, their locations and associated artefacts	122
Table 8.2	The main artefacts involved in police incident response sensemaking	128
Table 9.1	Key issues in multi-agency emergency response	140
Table 10.1	Responses to CDM questions from the various organisations, in relation to the risk assessment of having staff working inside the electricity substation	156

Preface

This book contains work that began under the Human Factors Integrated Defence Technology Centre (HFI-DTC), which ran from 2003 to 2012. Over this time, we were involved in a wide range of projects exploring the ways in which command and control was managed in contemporary military settings. This led us to think about situation awareness and how this was managed in technology-supported and networked groups. For us, a key issue became the matter of what preceded situation awareness: that moment when decision makers realise that there is a *situation* but are not yet sure how to formulate a response to it. We see in this moment the challenge of sensemaking, and whilst this was not explicitly addressed in the HFI-DTC work, this has played an important role in subsequent work.

In addition to the focus on military activity, we were fortunate that the HFI-DTC funded work into the study of emergency services, as a way of comparing practice across different organisations and as a way of exploring multi-agency response. This led directly to Richard McMaster's PhD thesis, which forms the backbone of this book. Over the course of his research, the focus of the thesis shifted from understanding the impact of digital radio systems on the organisational structure of the West Midlands Police Force to a broader appreciation of the police force as a socio-technical system. Pursuing this work led him to join Warwickshire Constabulary as a special constable, allowing him opportunities to participate in and observe a wide range of policing activities. This, combined with the HFI-DTC work on emergency response and multi-agency activity, resulted in a wealth of information that allows us to explore the ways in which *sense* is made during the initial stages, and over the course of responding to incidents.

The examples used in this book are all drawn from studies of emergency response, either from the perspective of an individual agency (the police force) or from multi-agency operations. The case study material used in the studies is taken from UK operations, observed over the past 10 years or so. This means that the content of the examples are specifically of a time and of a place. We were interested in how well the concepts of sensemaking could explain our observations, but are equally interested in how well the examples we present can help explain concepts of sensemaking. We are interested in that period of time during which an *incident* is initiated, say, in response to a call from a member of the public or from a report by an officer on patrol, and then defined, through the acquisition of additional information. To this end, the focus of the book is on the ways in which an organisation is able to grasp the situation to which it needs to respond and, in particular, the moment when this has been achieved. Hence, the title of the book is *grasping the moment* because we see this as that period of time during which the demands on an organisation's sensemaking capabilities will be most acute.

About the Authors

Dr. Richard McMaster earned his bachelor's and master's degrees in psychology at Sheffield University in 1999 and 2002, respectively. He then joined the University of Birmingham as a research associate who is funded by the UK Ministry of Defence's Defence Technology Centre for Human Factors Integration. Over a period of some nine years, he worked on a variety of projects that are related to military command and control and to the management of emergency responses. In addition to this work, he registered as a part-time PhD student, developing a thesis on the ways in which the emergency services made sense of routine and major incidents. This research was complemented by his role as a special constable (police volunteer) with the Warwickshire Police, which gave him a first-hand experience of police emergency response work. Dr. McMaster now works as a human factors specialist in the defence nuclear industry.

Professor Christopher Baber joined the University of Birmingham in 1990 as a lecturer on the MSc work design and ergonomics programme. Prior to this, he earned his PhD at the Applied Psychology Unit at Aston University (with a thesis on speech technology for control room operations, which was subsequently published by Ellis Horwood). His research focuses on human–computer interaction (particularly in terms of sensor-based interactions and wearable computers) and on distributed sensemaking (particularly in uncertain or complex domains, such as intelligence analysis). He has published more than 80 referenced journal papers, as well as around 400 conference papers. He is the author/co-author of six books (for Ashgate, CRC Press, Ellis Horwood, and Springer). In 2013, he was awarded the Sir Frederic Bartlett Medal by the Institute of Ergonomics and Human Factors for his contributions to research.

1
INTRODUCTION

In this chapter, we introduce the concept of sensemaking through examples of incidents involving phone calls to dispatch centres. The examples highlight the challenge of defining what is *made* during sensemaking and the importance of recognising the distinction between process and product. These points lead us to distinguish three types of sensemaking: (1) individual, (2) artefact-driven and (3) collaborative. We propose that the notion of distributed cognition provides a useful bridge between these types.

1.1 The Challenges of Sensemaking

On July 22, 2011, Anders Breivik set off a bomb in the city centre of Oslo, killing eight people before going on to massacre a further 69 people on Utøya Island. Norway was shaken, and a key focus of the Gjørv Commission, set up by the Norwegian government to review this tragedy, was the question of whether the killings could have been prevented. The Gjørv Commission's highly critical report lists a number of deficiencies in the response and led to the resignation of Norway's chief of police.

Of particular concern for this book was the manner in which information was obtained, shared and acted upon. A report in *the New York Times** noted, 'Ten minutes after the bomb detonated, a person gave them information about a man in a police uniform holding a pistol who was acting strangely. The person said he got into a gray van. He gave the license plate number… The person who took this call knew this was important. She brought this information to the operations center. This lay around for 20 minutes. Once it was passed on, it was not read until two hours later'.

---

* Lewis, M. August 13, 2012. In Norway, panel lists police faults in massacre. The *New York Times*. Retrieved September 9, 2014.

A translation of the transcript of this telephone conversation, recorded 10 minutes after the bomb explosion in Oslo, shows that the caller provided a great deal of information in a fairly unstructured and blurred account.

Dispatcher: Just tell me briefly what did you see?
Caller: I saw what I thought was a policeman, but then I was surprised. A man with a protective helmet and police clothes and a drawn pistol came up behind me, near the government building. I was surprised that he was walking alone, and I just followed him out of the corner of my eye. Then he got into a car, a grey van with the following license number …

The dispatcher recognised the importance of this information. However, the manner in which this was passed to the control room is somewhat peculiar. The dispatcher wrote the information on a note, marked it as *important* and took it to the control room. It was another 20 minutes before the note was noticed. Under the circumstances, one would expect the control room to be in a state of intense activity, and so the question of whether or not the note was noticed is less important than the more basic question of why this was used as a means of passing on important information. When the note was responded to by control room staff, there was a further delay of around two hours until the information was passed to control rooms in neighbouring counties.

It is always easy to judge situations with the benefit of hindsight, and our intention in using this example is not to add to the criticism that was made in the Gjørv Commission's report.* At an individual level, the caller recognised that the information was sufficiently important to pass it to the dispatcher, and the dispatcher recognised the importance and so passed it to the control room, which (after a period of time) recognised the importance of the information and passed it on to regional control rooms. This process of recognising the importance of information and seeking to share it characterises *sensemaking*.

Not only does this example demonstrate how sensemaking can break down; it also helps to introduce some key issues to be explored

---

* Büscher, M., Liegl, M., Perng, S., and Wood, L. (2014). How to follow the information? A study of informational mobilities in crises. *Sociologica: Italian Journal of Sociology*, issue 1: http://www.sociologica.mulino.it/doi/10.2383/77044.

in this book. For the individuals who are involved, the potential importance of the information would appear to be quite straightforward; under the circumstances of an attack on a civilian target, here is some information that could be relevant. However, this interpretation also highlights the need to look more deeply into *how* sensemaking is performed and how this performance moves through an organisation. Thus, sensemaking is about the process and performance of assigning relevance and meaning (or sense) to particular pieces of information, often in terms of a *bigger picture*, which needs to be explained, interpreted or acted upon. The broader context in which sensemaking takes place can create huge pressures on the sense makers who are involved in terms of high risk, high stakes and uncertainty (Alison and Crego 2008). To some extent, the explorations in this book also reference the question that was raised by Manning (1988): 'How does organized rationality interface with the variegated dilemmas and perplexities of human communication?' (p. xv). Manning (1988) expressed concern over the potential dangers of the 'pseudorationality ... of technocratic language' (p. xvi), and we are equally concerned that the ways in which *sense* is defined, structured and recorded in the various artefacts used in emergency services can become as much a means of distorting as a means of clarifying sense.

For Manning (1988), a key issue was the various ways in which rules are enacted in police response – these might be informal *rules* that reflect the particular concerns of police officers and call handlers (in terms of the acceptable ways of behaving on and off duty), or they might be formal rules that dictate how information is recorded, shared and acted upon. Any discussion of sensemaking needs to be cognisant of this patchwork of rules that influence the space in which information is interpreted and the ways in which different *readings* of the same information can vary, depending on the audience. For example, the sense that might be informally discussed over a drink between colleagues after work might differ from the sense that is presented in response to a question from a barrister in court. This book is not directly about these differences, but as we present different examples, the readers are invited to consider how the context in which sensemaking is being performed can influence what sense is being permitted or allowed to develop, and how some sense might be constrained by context.

In addition to sensemaking being a process, the Breivik example also highlights that this process can be performed at both an individual and organisational level. When performed at an organisational level, it is likely that there need to be some means of passing the results, or product, of the sensemaking process to other individuals. In this example, that product was a handwritten note that was taken to the control room. One might feel that this was the central failing of this incident and that everything would have been improved with the introduction of some form of (digital) technology. The Gjørv Commission's report speculated on whether the forwarding of this information could have prevented the subsequent attacks.

We shall argue in this book that technology is *not* the panacea for improving sensemaking. For instance, *if* the dispatcher was able to directly forward the content of the report of the caller's information, this would not guarantee that it would be immediately attended to (it could have been held in a queue of other messages or emails), nor would simply requiring a particular message to be tagged *urgent* necessarily have led to a faster response. Remember that this information was only a small part of the wealth of information coming into the control room just after a bomb had exploded in Oslo, leading to a great deal of stress on the system. The job of the dispatcher was to triage incoming information and pass on important material. In this example, it is clear that this is what the dispatcher had done. This raises several questions for this book. For example, how do dispatchers (or call handlers) make sense of the calls that they receive, and how does this sense inform the response that is made to the call? How does this response involve the refining of the sense into an understanding of the incident in order to ensure proportionate and appropriate response? How does this understanding of the incident become shared with other parties during the course of the response? How do the products and technologies available to people in the organisation help (or hinder) their ability to engage in sensemaking?

In this book, we examine sensemaking during routine response and in multi-agency major incidents in the United Kingdom (UK). The primary focus of attention is on police response activity. The duties of the emergency services include far more than just responding to incidents and emergencies, but these other duties will receive less attention in this book. In the context of policing, Ormerod et al. (2008)

note that *'A critical, though as yet poorly understood, aspect of expertise in criminal investigation is the set of knowledge and skills associated with "sensemaking", whereby an investigator uses available information to construct an understanding of the "to-be-investigated" or ongoing incident'* (p. 81). Whilst we are not directly concerned with the activities that are related to subsequent investigation, we are interested in the short window of time between a call being received and a response being initiated to an incident.

1.1.1 When Does Sensemaking Happen?

It might seem odd to ask the question of *when* sensemaking happens because the obvious answer would be to say that sensemaking happens whenever it needs to. The problem with such an offhand response is that it does not reflect the idea that there might be situations and circumstances when sensemaking would be more useful or more likely to happen. The Breivik example we have used to introduce this chapter suggests that sensemaking happens when someone is presented with a collection of material that needs to be put into some form of order. As Weick (1995, p. 9) noted, sense '*... must be constructed from the materials of problematic situations that are puzzling, troubling and uncertain*'. This introduces three characteristics of situations in which people might engage in sensemaking, i.e. (1) puzzling, (2) troubling and (3) uncertain. Applied to the conversation between the caller and the dispatcher in Section 1.1, we can see that the caller is responding to a puzzle – 'I saw what I thought was a policeman, but then I was surprised' – and is seeking to convey this sense of uncertainty to the dispatcher in order to elicit a response, i.e. for the dispatcher to indicate that the information would be useful to the investigation. What we can also assume here is that the situation is not only puzzling but also threatening (there has been an explosion), complex (the cause of the explosion is still being ascertained) and with high information load (there will be many calls coming into the control centre).

From this discussion, sensemaking occurs in response to the novel, the uncertain and the complex. Emergency response is often characterised by its *un-ness*, i.e. unexpected, unprecedented and unmanageable (Hewitt 1983). Thus, this would seem to be an obvious context in which to explore sensemaking. However, it is also worth noting that a key role of the dispatcher is to triage incoming information in such a way as to determine whether the situation requires sensemaking or whether it can be labelled as a *routine* event that can be handled with standard operating procedures and minimal resources. In this case, the initial interpretation of the situation will involve defining its meaning, i.e. what type of situation is this? And what type of response does it require? Such meaning-making characterises an overarching aim of sensemaking. As the situation unfolds, then standard operating procedures may no longer be appropriate, and a new sense might need to be determined. This suggests two things. First, sensemaking arises in response to situational characteristics. This implies that it is important to be able to appreciate when to engage in sensemaking. Second, sensemaking is a continuous process of gathering information, creating meaning and determining an appropriate response.

1.1.2 What Is Made in Sensemaking?

In the preceding discussion, a distinction was drawn between the process of making sense of information and the resulting product that is used to record and share the outcome of this process. This highlights a challenge that is posed by the concept of sensemaking. It is also worth noting that, for some writers, the phrase is sense-making (note the hyphen). The word *making* implies that something is constructed, perhaps from the pieces of information that an individual receives from his or her environment and perhaps from combining these pieces of information with what the individual already knows. Thus, there is clearly a cognitive activity that is involved in recognising a need to engage in sensemaking, in selecting and combining the information, in evaluating the outcome of this activity and in communicating this outcome. We prefer to use the contracted word *sensemaking* to emphasise that this is a process of doing sensemaking rather than an activity of making something that is called sense. The notion that sensemaking results in a thing called sense is problematic because it is not obvious what this outcome might be. Of course, one could simply say that what is made is sense; but this tautological argument does not help, particularly because it is not clear what sense might be or how one knows that sense has been made. For now, we will note that sense can take at least two forms.

In the first form, sense is a subjective impression (by the sense maker) that some impasse has been overcome, that some gap in the understanding of the environment has been filled or that some feeling of uncertainty has been resolved. We see this as an activity in which a blurred view of the environment or situation is brought into sharper focus.

The second form of sense is the product in which the description of a situation is created and shared. So, in the Breivik example, sense is the message on the note (together with the record of the call that the dispatcher probably completed using the call-handling software). This second form of sense is what is passed through an organisation, possibly forming the basis for subsequent sensemaking activity by other individuals in other parts of the organisation. Weick (1995) describes sensemaking as *'a developing set of ideas with explanatory possibilities, rather than a body of knowledge'*, which goes some way to explain the variety of interpretations. This could be seen not simply in terms of the artefacts that an organisation uses to record and share information but also as a way in which *common ground* can be established as the basis for communication between the individuals in that organisation.

This comparison of two forms of sense highlights a key challenge in understanding sensemaking. On the one hand, sense is the subjective impression that individuals might form when combining pieces of information into a holistic view of a situation. As such, there is the potential for this subjective impression to take the form of a belief (and hence the potential for idiosyncracies in interpretation). On the other hand, sense is the objective marshalling of facts to reach a definition or conclusion (which can be recorded and shared with colleagues). The reality lies between these two extremes and, ideally, at a point at which individuals can articulate, explain, justify and defend their sense on the basis of the evidence that is available to them. One also expects the differences between expert and novice sense makers to lie between these extremes, perhaps with the expert ones showing not only a greater ability in articulating and defending their sense, on the one hand, but also a greater capability to *feel* that the situation warrants sensemaking. Noting these extremes can help us to appreciate that the technologies designed to *support* sensemaking may not always be effective because they will impose one form of sense on activity in a manner that might not be appropriate to reasoning and interpretation being performed by the individuals engaging in sensemaking. It can also help to appreciate that just because someone *knows*, or rather believes, that his or her sense is incontrovertible does not mean that it should not be challenged if someone else has access to a different evidence or can view the evidence in a different way. This process of challenging is not meant to imply an argument or a contradiction, so much as to reflect the normal processes by which people make sense through sharing information in conversations. In other words, in this context, *challenging* refers to continually checking subjective impressions and verifying objective facts. In the next section, we consider the ways in which a conversation operates on more than one level to reach some common ground among the people who are involved in that conversation.

1.2 Common Ground in Conversations

The following extract is from a call that was presented by Whalen and Zimmerman (1990) and formed the basis of a discussion in a paper by the authors (Baber et al. 2006).

Dispatcher: .hh Midcity emergency
Caller: .hh Yeah uh (m) I'd like tuh: -report (02) something weir:d that happen:ed abou:t (0.5) uh (m) five minutes ago, 'n front of our apartment building?

The caller has some information regarding 'something weir:d' that might require a police response. Rather than directly stating the facts, the caller is seeking to pique the dispatcher's interest by offering an opportunity to engage in a conversation. The rising tone at the end of the caller's contribution (as indicated by the '?' in the extract) suggests that the initial part of the call is partly to ensure that the dispatcher will acknowledge that this is a legitimate enquiry. The fact that the time and location are given so imprecisely could be seen as invitations to the dispatcher to ask more specific questions.

The description of the call relating to Breivik begins with the observation that the caller provided a *great deal of information in a fairly unstructured fashion and blurred account*, and the example from Whalen and Zimmerman (1990) shows the caller presenting an imprecise and hesitant opening to their call. This unimpeded speech of the caller can often help them to create their own sense of what they are calling about and help the dispatcher to spot key pieces of information in the person's speech (or, indeed, in the tone of their voice) that can help determine the nature of the incident and the type of response that might be required. For example, if it appears that the incident is serious and ongoing, the dispatcher might immediately raise a response and then seek further information from the caller.

As Baber et al. (2006) point out, the dispatcher will seek to enter details from the call into an incident log. The incident log provides a record of the call and information that can be shared with other people (either during the course of the response or during post-incident investigation if required). However, rather than taking a verbatim account of the caller's information, the dispatcher will translate this information into a format that is more suited to the structure of the incident log.

The caller and the dispatcher seek to establish some form of common ground. The concept of common ground, as a means of supporting communication between people, has been developed by Herbert Clark and his colleagues (e.g. Clark and Marshall 1981; Clark and Carlson 1982; Clark et al. 1983; Clark 1992, p. 81). In this concept, common ground is *'the mutual knowledge, beliefs, and assumptions shared by the speaker and addressees'* (Clark 1991, p. 247). An important question at this point is the extent to which the word *mutual* implies identical and complete – there might be an assumption that both the speaker and addressee would need exactly the same information in order for a conversation to proceed, or that the purpose of the conversation was for the information that was held by the addressee to become exactly the same as that which was held by the speaker. Clark et al. (1983) point out that a *traditional* view of conversation assumes that the speaker simply needs to present utterances that can be understood, and the hearer simply needs to understand these. However, in conversation, the speaker and the hearer need to do more than this: they need to *'establish the mutual belief that the addressees have understood what is uttered, to establish what the speaker meant as common ground'* (p. 124). From this perspective, the caller's speech is a way of getting the dispatcher to determine the most appropriate level of detail to provide, the best means through which to express this detail and the order that the detail ought to be provided.

Parts of the conversation might proceed through the hearer providing vocal responses to the speaker (such as, mm, uh huh, yeah, or 'k), but these do not always mean that the speaker's utterance has been understood – these responses could, instead, mean 'go on with what you are saying, and I will try to make sense when I hear more'. The common ground, therefore, is as much a matter of indicating whether the conversation can proceed as whether both parties have the same information. By implication, common ground is partly about managing a conversation rather than being just about each partner obtaining the same information. Our interviews and discussions with dispatchers suggest that many of them will allow the caller to speak without interruption before asking questions.

Clark (1996) suggests that common ground can be considered as a distributed representation. This notion of distributed representation becomes even more significant when we consider the role of technology in sensemaking. In the example that was used in Section 1.1, information was received from a dispatcher, logged on to a computer and written on a note.

Clark's concept of common ground proposes that people draw on three sources of information:

1. *Perceptual evidence* (the experience to which people have access)
2. *Linguistic evidence* (the words that people are hearing)
3. *Community evidence* (knowledge that people might believe is shared within a given community, perhaps as the result of training or enculturation)

In the examples considered so far, the caller and the dispatcher do not have the same perceptual evidence. (The dispatcher is far removed from the scene that the caller is witnessing or recalling.) Thus, part of the conversation is aimed at translating the caller's perceptual evidence into a format that can be operationally useful. At one level, this might mean extracting only those aspects of the caller's account that can be entered into the incident log. This might mean that the dispatcher would simply ask questions to prompt for information in each part of the log. For incidents that do not require an immediate response (say, a resident reporting a new piece of graffiti or reporting a crime that is discovered after a weekend away), the dispatcher might work through the form as it is laid out. For other incidents, the dispatcher might jump around the form or seek information that cannot be easily entered on the form (because he or she is dealing with an unusual incident or with a distraught caller or with an incident requiring immediate response). This suggests that some conversations could be considered as driven by the need to complete the incident log, whereas others are driven by the need to determine the nature of the incident (and the corresponding response). Having said this, the incident log can also become the first view of the incident that the responding officer might have, and he or she would base his or her expectations of the incident on the content of this log.

In terms of linguistic evidence, a key role of the dispatcher is to translate the words of the caller into the terminology that is used by the police. In part, this is a matter of replacing some everyday words with police acronyms and jargon, and, in part, this is a matter of seeking to ensure that the information provided is as clear and unambiguous as possible to allow a decision to be made on the nature of the incident and the response that is required. As this information is passed through the response chain, more material could be added, or other sources of information could be consulted. For example, as the response develops into an investigation, police officers might seek eyewitness statements and then seek to corroborate or challenge the original call or other statements.

In terms of community knowledge, one might find the caller's seeking to provide information in a manner that he or she believe fits the community knowledge of the police. For example, when reporting a car's registration number, the caller might use the International Civil Aviation Organization alphabet of A = alpha, B = bravo, C = Charlie, etc., because he or she believes that this is how to report the letters of the alphabet to a police officer. The translation of caller terminology to police acronyms and jargon is a further example of the way in which the incident becomes translated into police terms from the caller's terms. These forms of *translation* become a form of sensemaking, i.e. through which a given vocabulary is used to record a report that is presented in a different vocabulary.

1.2.1 Sensemaking and Common Ground

We noted in Section 1.2 that common ground is a matter of ensuring sufficient understanding to allow a conversation to progress (rather

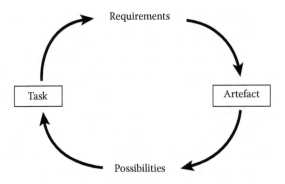

Figure 1.1 Task–artefact cycle.

than necessarily to ensure full, mutual knowledge). In other words, it is less important for the two people in a conversation to completely know what each knows, so much as to have sufficient overlap to enable the conversation to move forward. Taking this a step further, this suggests that common ground involves determining the most appropriate actions to perform and, potentially, the most important actors to involve in sensemaking. It could also mean that a key aspect of a conversation is the way in which each speaker seeks to create opportunities for the other to make a contribution. In terms of sensemaking, this highlights a pragmatic aspect of behaviour in which what is made is not only sense as a semantic (meaning-laden) product but also as the opportunity to amend or modify this product. We note that sometimes, this product could take a physical form, e.g. in terms of entering information into the various artefacts that organisations require people to use. This, in turn, creates new opportunities (or indeed constraints) for action. As Carroll et al.'s (1991) *task–artefact cycle* (Figure 1.1) illustrates, acting on an artefact requires the performance of a task, which leads to a change in the state of that artefact, which leads to the definition of new tasks in response to the changed state.

1.3 Three Types of Sensemaking

The preceding discussion helps illustrate a point that we will be making throughout this book. Sensemaking is not reducible to a single activity or a single product. Rather, there are combinations of activity and products that illustrate different types of sensemaking. From this distinction between activity and product, for instance, we can propose at least three types of sensemaking.

1.3.1 Individual Sensemaking

The first type of sensemaking involves an activity in which an individual collects, collates and sifts information (making) in order to produce a coherent and consistent account (sense) of a given situation or phenomenon. This represents sensemaking as a cognitive activity, closely allied to problem-solving or situation awareness, in which the experience and expertise of the person shapes the actions that he or she performs. In highly dynamic or uncertain situations, the person might need to continually adjust the manner in which information is sought, collated and filtered and the manner in which the resulting sense is tracked. This represents a higher-level, or metacognitive, activity. The person is not only performing the cognitive activity of collecting, collating and sifting information but also monitoring and adjusting the strategies in which these actions are performed.

1.3.2 Artefact-Driven Sensemaking

The second type involves an artefact being used to structure the activity. For example, a form could be used to structure an interview, and the interviewer could simply ask each question as it appears on the form and record the answers as they are given. Rather than being an active process of collection, collation and sifting of information, the activity could be a passive process of following the prompts on the form. This represents sensemaking as an artefact-driven activity in which the structure of the form creates the actions that the person performs; thus, the collection of information is determined largely by the design of the artefact. In this way, the form reflects the rules (Manning 1988) that the organisation uses to define the importance of information. The design of the artefact is, in turn, influenced by the ways in which an organisation stores and shares information. For instance, the form could be the user interface to a database in which the call information is held or could be a screen that can be forwarded to other people, e.g. in the form of a call log.

It is proposed that this artefact-driven sensemaking can be efficient for dealing with routine incidents (in which the same information is captured on each occasion), but when new information contradicts or undermines the current version of the sense of the situation, the use of the artefact might not be sufficient to keep track of these changes, or as the complexity of the incident might be ignored in favour of only focusing on the information that can be recorded using the artefact. This represents artefact-constrained activity in which the limitations of the technology restrict the activity that the person can perform. We argue that artefacts can constrain sensemaking when they are designed to support one type of activity (such as collecting information for managing and auditing types of response) but are introduced to situations that might involve another type of activity (such as ensuring that an appropriate response is made). Often, these activities converge (and it is likely that the original design of the artefact was motivated by a particular view of how response is managed), but problems arise when one of the activities diverges from this original view.

1.3.3 Collaborative Sensemaking

Sensemaking can be a social activity in which several people (or people and technological artefacts) contribute to the activity of making sense, either through sharing information or through collaborating on the interpretation and analysis of the information. In this type, it is not only important to be able to cooperate and collaborate on the activity of collecting, collating and sharing information, but it is also (and we suggest equally) important to understand who needs to be involved in this collaboration. This suggests that sensemaking involves knowing with whom to share information (and where to obtain information). This notion is akin to Wenger's (1998) notion of transactive memory and suggests that, rather than collaborative sensemaking simply being individual sensemaking writ large, it involves additional activity and demands. Thus, it is important to recognise that sensemaking is seldom performed by one individual in isolation. Rather, sensemaking is a social activity that is performed in collaboration with other actors (both human and automated).

1.4 Macrocognition and Sensemaking

We noted in Section 1.3.3 that, in addition to the cognitive activity of sensemaking, it is also important for people to be able to monitor the strategies that they use. This higher-level activity is reflected in the macrocognitive framework, summarised in Figure 1.2. Sensemaking is thought to initiate and influence adaptive planning and decision making (Lin and Klein 2008).

This suggests that sensemaking is not only about cognitive processes that are involved in handling information but also about processes that guide and manage this cognition, i.e. metacognition. To help describe sensemaking at a higher level, Klein et al. (2007) state nine characteristics of sensemaking, which set an agenda for this book:

1. Sensemaking is the simultaneous process of fitting data into a frame and fitting a frame around the data.
2. Data elements are inferred, using the frame; different people may derive different data elements from a situation.
3. The frame is inferred from a few key anchors, and this frame is used to search for more data elements.
4. The inferences used in sensemaking rely on abductive reasoning (i.e. the most plausible explanation) as well as logical deduction.
5. Sensemaking usually ceases when the data and frame are brought into congruence.

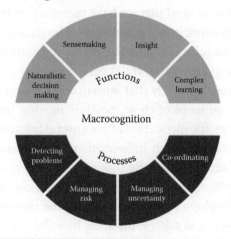

Figure 1.2 Macrocognitive framework. (© Macrocognition LLC, 2010. Available at http://macrocognition.com.)

6. Experts reason the same way as novices but have a richer repertoire of frames.
7. Sensemaking is used to achieve a functional understanding – what to do in a situation – as well as an abstract understanding.
8. People primarily rely on just-in-time mental models: '… *constructed from fragments … In complex and open systems, a comprehensive mental model is unrealistic*' (Klein et al. 2007, p. 151).
9. Sensemaking takes different forms, each with its own dynamics (Klein et al. 2007, p. 120).

1.5 Distributed Cognition as a Unifying Concept for Sensemaking

Whilst one might expect that all sensemaking follows either the first or third type (on the assumption that sensemaking is an active, cognitive activity that is performed by experts who are knowledgeable in their domain of operation), there are plenty of examples of artefact-driven sensemaking, especially in work domains that are heavily proceduralised or that have high levels of automated support, such as computer-based forms that require filling. In this way, sensemaking represents a form of distributed cognition that will be explored in this book. The suggestion that there is a clear link between sensemaking and distributed cognition not only clearly arises from our proposal that sensemaking can be artefact-driven but also links to the interpretation of common ground that we have presented in this chapter. We believe that such a link is underexplored in the sensemaking literature and provides a useful way of looking at this phenomenon.

We are not suggesting that the types are discrete and distinct from each other; rather, it is likely that sensemaking will operate in a combination of two or more of these, possibly moving from one type to another. However, the distinction offers a useful framework to compare the different ways in which individuals and organisations approach the problem of sensemaking. By linking the three types through the concept of distributed cognition, we illustrate how combinations of people and technology make sense during the management of incidents. This allows us to ask, for instance, how the artefacts that people use in incident response support and constrain their activity. We are not arguing that *constraint* is necessarily a negative aspect of the role of these products; for example, having a clearly defined structure for reporting information and a clearly defined vocabulary for sharing the information *could* be useful to support common ground. However, it is important to recognise how such constraint operates and to determine when constraint might need to be either relaxed or replaced. Furthermore, we see constraint as arising from the command and control (C2) structures in which individuals operate, and we show how these structures can also have positive and negative consequences for sensemaking. It is our contention that the ways in which people circumvent constraints (perhaps through skipping over the sections of a form that they do not feel to be relevant, or by using additional products, such as scraps of paper to write notes before completing a form) reflect the knowledge and experience of the people who are involved as well as the variability of the situations that they face.

1.6 Graphically Representing Distributed Cognition

Whilst much of the literature addressing either sensemaking or distributed cognition tends to present detailed verbal accounts of systems and their operations, it can often be informative to supplement such accounts with a visual summary. In this book, we use an approach that was developed during observations of a military exercise. Details of the exercise and the analysis conducted can be found in McMaster and Baber (2005). In this section, we present a short example of the graphical representation.

During the exercise, it was clear that a great deal of the cognitive activity was spread across several actors and the artefacts that they were using, and there was seldom an instance in which a single individual could be said to be co-ordinating all of this activity. This is typical of the fast-moving environment in which new information comes into a military command post and needs to be processed in order to determine an appropriate course of action. Often, the information is received by separate individuals (either because the information is fed directly to them from a source with whom they are in contact, or because the information is routed to them because it requires their expertise to interpret). Individuals work on the information at hand and then share observations with colleagues until a general consensus is reached. This consensus could involve arbitration by a commander, e.g. the commanding officer would receive briefings from all the individuals and then decide on the course of action, or it could involve mutual agreement amongst all individuals, as in the example in this section.

In this example, a Joint Engagement Working Group (JEWG) was tasked with reducing a list of potential targets in the exercise to a workable number, which could then be prioritised and submitted to the commander. Individual members of the group gathered information on nominated targets and made their own estimate of priority. Information was shared in the form of *target sheets*, i.e. spreadsheets with each column representing a target attribute, such as location. As the meeting progressed, blank cells in the spreadsheet are either filled in (by the members of the group who have that information) or form the basis for a request for information from other sources. We have noted in Section 1.3.2 that the role that an artefact plays changes during the course of the activity, i.e. from a cue for information gathering, to a form of information checking, to a cue for requests for information, to a means of agreeing priority, and this was certainly the case for the target list. Not only is the target list the format in which information search is structured but also the medium through which the information is shared, and, at the end of the activity, it provides a record of the *votes* of the team members (in terms of priorities). Thus, the target list becomes not only the storage of information but also a means of co-ordinating and managing activity.

Table 1.1 gives a list of features that can be identified for each operation within the activity and relates to the wider cognitive actions that are taking place at that point in the process.

Figure 1.3 shows a single operation from the target selection process. Such a diagrammatic representation illustrates the key transformations of information (e.g. from one modality or storage medium to

Table 1.1 Distributed Cognition Features of Each Activity within a Task

FEATURE	DESCRIPTION
Agents	Who is involved (people/artefacts)
Activity	The purpose of the operation
Modality	Information state (verbal, text, etc.)
Form	Language style, abbreviations, etc.
Transmission	How information is shared
Transformation	How information is acted upon
Storage	How information is retained
Resource for action	Actions cued by representation

INTRODUCTION

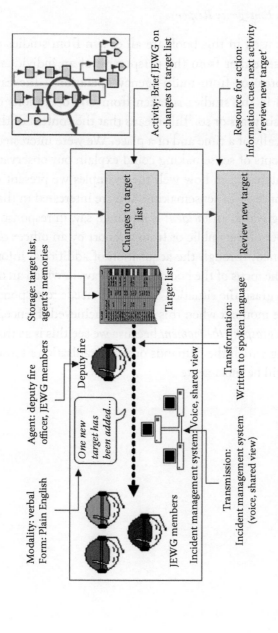

Figure 1.3 Graphically representing distributed cognition in the target review process.

another). The annotations on the flow charts indicate the changing nature of the role of target lists during the activity.

1.7 A Note on *Emergency Response*

The examples used in this book are all drawn from studies of emergency response, either from the perspective of an individual agency (the police force) or from multi-agency operations. The case study material used in the studies is taken from UK operations observed over the past six years or so. This means that the content of the examples is specifically of a time and of a place. We were interested in how well the concepts of sensemaking could explain our observations but are equally interested in how well the examples we present can help explain the concepts of sensemaking. We are interested in that period of time during which an *incident* is initiated, say, in response to a call from a member of the public or from a report by an officer on patrol, and then defined, through the acquisition of additional information. To this end, the focus of the book is on the ways in which an organisation is able to grasp the situation to which it needs to respond and, in particular, the moment when this has been achieved. Hence, the title of the book is *grasping the moment* because we see this is as that period of time during which the demands on an organisation's sensemaking capabilities will be most acute.

2
INDIVIDUAL SENSEMAKING

In this chapter, we consider sensemaking as a form of individual cognition. Working from the naturalistic decision-making (NDM) perspective, we consider sensemaking as a form of recognition-primed decision (RPD) making and in terms of the data/frame approaches. In these approaches, the individual will extract information (*data*) from a given situation and seek to apply a *frame*, which can explain the connections between these data. The challenge is to consider how data are selected and how frames are applied.

2.1 Introduction

In Chapter 1, we proposed three types of sensemaking. This chapter will elaborate on the first of these types. Sensemaking is, in the words of Ascoli et al. (2014, p. 82), '... *a temporally extended inference task involving multiple cycles of information foraging, evaluation and judgement*'. In other words, sensemaking involves an active response to a situation, taking in information from the world and structuring it in order to inform the choice of response. In this way, sensemaking requires an individual to engage in cognitive activity in which information is collected, collated and sifted (*making*) in order to produce a coherent and consistent account (*sense*). In this chapter, we consider how this activity might be performed, focussing in particular on ideas that are developed in the field of NDM. Central to this view is the notion that sensemaking is carried out by individual experts weaving together different strands of activity (Figure 2.1).

One perspective on this challenge of establishing what is happening sees sensemaking as a process for reducing discord, or the *gap*, between one's expectations and the development of events (Bjørkeng 2010). The suggestion is that, in any situation, we apply

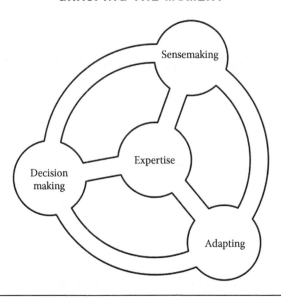

Figure 2.1 Strands of expert activity in the field.

prior knowledge and experiences in order to interpret the current state of affairs. When our prior knowledge and experience are not sufficient to do this, we experience a sense of discord or a gap in our explanation (Dervin 2003).

It is possible that features of a situation might arise that *ought* to be responded to as an opportunity for sensemaking, but the person might miss this opportunity. The likely consequence of this is that the act of *seeing the gap* becomes delayed temporarily and returns with a growing discord later. Thus, the experienced incident responder might approach a situation as if it contained a sensemaking opportunity (seeking to identify discord rather than apply his or her experience), whereas the rookie might seek patterns of information that he or she recognises and can label in terms of prior experience. This also implies that approaching a situation through the lens of an artefact, i.e. a form to complete, might result in the sensemaking opportunity being distorted because the gap becomes defined by the fields in the form rather than the features of the situation.

To summarise so far, sensemaking involves recognising the gap between what you know and what you need to know in order to understand a situation. Landgren (2004, 2005b) describes this as the progressive clarification of a situation in an iterative process of *committed interpretation*, through which an individual's behaviour (actions) influences further sensemaking (and further actions):

> ... *committed action creates the context for interpretation by narrowing the actors' focus to a subset of cues in the available information that suggest reasonable justification of those actions.* (Landgren 2004, p. 91)

This distinction between *committed interpretation* and *committed action* raises the contrast between collecting, collating and sifting information (i.e. acting on information) and selecting an action to perform (i.e. acting on the situation). This implies that sensemaking is not simply a matter of filling the gap but also (and more importantly) of recognising and acting upon the gap. Put another way, sensemaking begins with some knowledge arising from active exploration of the situation. This leads to what Baron and Misovich (1999) term *knowledge by acquaintance*, which involves responding to particular features of the situation. In order to explore this knowledge in detail, or share it with other people, there is the need to shift from knowledge by acquaintance to *knowledge by description*, which, in turn, evokes the use of prior experience.

2.2 Representing Prior Experience

Prior experience and knowledge are an important aspect of sensemaking. This raises the question of how the sense maker represents, accesses or makes use of such information. A dominant view in the cognitive sciences proposes that experience and knowledge are maintained in some form of knowledge structure, which, in turn, is maintained in some form of long-term memory. One of the most popular concepts used to describe such knowledge structures is a *schema* (plural schemata), which literally means form or shape (from the Greek word σχῆμα, schema). The suggestion is that experience and knowledge become shaped into a form that can be reused. Whilst it feels intuitively useful to see schemata as knowledge structures that are built from past experience (Bartlett 1932), and as a way of thinking about how we think, the concept has proven difficult to pin down. Indeed, a broad definition of schema was offered by Taylor and Crocker (1981) who suggest that it can be useful in at least seven different functions, viz.,

1. Providing a structure against which experience is mapped
2. Guiding information encoding and retrieval
3. Affecting information-processing efficiency
4. Guiding the filling of gaps in the available information
5. Acting as a template for problem-solving
6. Enabling the evaluation of experience
7. Enabling anticipations of future states, goal setting, planning and action

The problem with such a breadth of function is that this rather weakens the usefulness of the concept itself; from this list, one could be forgiven for seeing a schema as a general-purpose support for thinking rather than a structure for knowledge. What the concept of schema does offer is a shorthand description of the everyday experience that we have when the features of a situation chime with our experiences and knowledge in a way that allows us to rapidly and easily interpret that situation. This suggests that a more useful approach to appreciating the role of schema is not to ask what they are but rather to ask how they are used.

2.2.1 RPD Making

Expertise can be related to the accumulated knowledge and experience of particular types of situation. The expert is able to extract key features of the situation and relate these to their experience and knowledge. This process has been captured by the RPD model (Figure 2.2), which was developed through interviews with experts, initially experienced fire service commanders (Klein and Crandall 1996). Subsequently, support for the model was demonstrated through studies in a range of uncertain, dynamic and time-pressured contexts (Klein and Crandall 1996; Klein 2011). The central argument of the RPD model is that experienced decision makers identify a suitable course of action through two cognitive processes: (1) situation assessment and (2) mental simulation (Lipshitz 1993; Klein and Crandall 1996); personnel identify the essential features of a situation, which

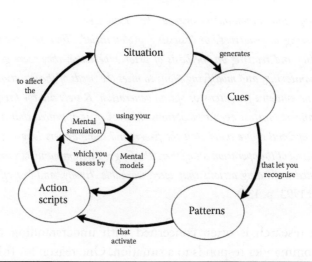

Figure 2.2 RPD model.

they relate to achievable goals in order to define a credible course of action (Klein and Crandall 1996). The course of action identified is then evaluated through mental simulation, which may lead to modification or rejection of the option (Klein 1993).

It is a moot point as to whether mental simulation is a necessary aspect of RPD. Indeed, experienced decision makers seldom generate multiple options and largely manage to identify a satisfactory course of action in the first instance (Klein 1993; Klein and Crandall 1996). The subsequent evaluation process is considered optional, depending on the closeness of fit between the current situation and previous similar incidents (Cohen 1993; Klein and Crandall 1996):

> ... *decision makers often have to cope with high-stakes decisions under time pressure where more than one plausible option does exist, but the decision makers use their experience to immediately identify the typical reaction. If they cannot see any negative consequence to adopting that action, they proceed with it, not bothering to generate additional options or to systematically compare alternatives.* (Schraagen et al. 2008, p. 4)

Through RPD, experienced personnel are able to make rapid, effective decisions in stressful, time-pressured environments (Lipshitz 1993; Klein and Crandall 1996):

> The fireground commanders argued that they were not 'making choices', 'considering alternatives', or 'assessing probabilities'. They saw themselves as acting and reacting on the basis of prior experience; they were generating, monitoring, and modifying plans to meet the needs of the situations. We found no evidence for extensive option generation. Rarely did the fireground commanders contrast even two options. Moreover, it appeared that a search for an optimal choice could stall the fireground commanders long enough to lose control of the operation altogether. The fireground commanders were more interested in finding actions that were workable, timely, and cost effective.
> (Klein 1993, p. 139)

RPD research is often concerned with understanding how the expert commander responds to a situation. One reason for this is that accounts tend to be retrospective, based on descriptions that are given by the commanders, and present the incident as moments of *decisive action*. In contrast, this book begins the analysis from the moment that an incident is detected, seeks to explore how the interpretation of the situation is generated and takes the incident response system (agents and artefacts) as the unit of analysis.

Given the tendency for experienced personnel to produce only a single course of action in the majority of instances (which may not even be evaluated), this raises the question of what contribution *deciding* makes to RPD. Klein (1993, 2011) acknowledges that sensemaking and decision making are related (with the former cueing the latter), but he maintains that they are distinct processes and that sensemaking does not wholly determine decision making. Sensemaking aims to establish *what is happening*, whereas decision making addresses the question, *what shall we do about it?* (Landgren 2005a). For this book, RPD provides a useful approach to understanding individual sensemaking, i.e. in terms of collecting, collating and sifting information.

It is worth asking, when we speak of individual sensemaking, whether there are differences between the different *individuals* who are involved in the incident. In Chapter 1, we presented examples where a caller spoke with a call handler. Each party to the call was seen as contributing to the production of sense (either by providing information or by structuring the conversation to develop common ground). This implied that sense arose out of conversation. In contrast, RPD has implied a privileged role for senior decision makers, that is, expert commanders '... *are forced to depend on the dots* [information] *and analyses that people at lower levels, with less expertise, are using*' (Klein 2011, p. 193). Whilst we support the view that sensemaking *can* take place at the individual level, we do not necessarily support Klein et al.'s (2006b) assertion that sensemaking is primarily the preserve of the individual commander. Where the opportunity permits, agents within the command and control system will collaborate to make sense of a situation, and this collaboration seems even more necessary for complex incidents. Indeed, the *individual commander* view, in incidents where multiple agencies work in the same physical space, can cause problems in terms of how the incident is framed. This problem could be obscured by a particularly dominant or charismatic commander at the centre of the incident, but if he or she does not fully understand or appreciate the manner in which other agencies operate, then mistakes and misunderstandings are inevitable. Before we explore this, it is important to consider how the sense makers detect and respond to gaps in the information that is available in a situation and their experience and knowledge.

2.3 Seeing the Gaps: The Data-Frame Model

For RPD, situation assessment forms a crucial role in the whole process:

> ... *making decisions in realistic settings is a process of constructing and revising situation representations as much as (if not more than) a process of evaluating the merits of potential courses of action.* (Lipshitz 1993, p. 133)

The process of recognizing a situation and matching it to a schema is complex – potentially involving '*a series of transformations and retransformations of the problem until the expert finally "knows" how to solve it*' (Cohen 1993, p. 67). We feel that the cognitive demands that this *series of transformations* implies seems to contradict the rapid response that Klein (1993) identifies. Situation assessment is a major factor that distinguishes decision-making performance between experienced and inexperienced personnel. Situation assessment can often involve simply *seeing* the features of a situation, and this is as true of call handlers as it is of first responders. In his discussion of insight, Klein (2013) reports an incident in which a relatively junior police officer sees a car and decides that it is stolen: *'As they waited for the light to change, the younger cop glanced at the fancy BMW in front of them. The driver took a long drag on his cigarette, took it off his mouth, and flicked the ashes onto the upholstery. "Did you see that? He just ashed his car," the younger cop exclaimed'* (p. 4). In this example, the *insight* that the *younger cop* had was that people do not normally mistreat a new car (not if it is theirs, not if they are borrowing a friend's car or not if they are delivering it to a customer). So, the behaviour was not easy to explain in terms of *normal* and created a gap that was filled by the supposition that the car was stolen (which it was).

Klein et al. (2007) view sensemaking as a process that takes place in dynamic situations in order to guide not only understanding but also action:

> *The active exploration of an environment, conducted for a purpose, reminds us that sensemaking is an active process and not the passive receipt and combination of messages.* (p. 118)

Klein et al.'s (2006a,b) data-frame model treats frames in a similar manner to schemas,* i.e. that they take the form of a retrospective narrative account, based on expertise and experience. For Klein et al. (2006a), the process of sensemaking involves the recognition and fitting of data into an appropriate frame, which then guides further data collection, and influences the filtering of data that are viewed as relevant to the situation (Figure 2.3). These processes of frame construction/modification and frame-defined data collection are thought to occur in parallel (Klein et al. 2005). Klein et al. (2006a) note that frames themselves change with the acquisition of data and that frames shape (transform) the data that they encapsulate.

From this perspective, problem detection is seen as part of sensemaking and is characterised as a hunch that *'the way events are being interpreted is incomplete and perhaps incorrect'* (Klein et al. 2005, p. 20), i.e. the sense maker encounters a gap in understanding. The hunch that motivates questioning a frame is based on the available data and

---

* Klein et al. (2007) describe the frame as the synthesis of concepts that are proposed by earlier researchers, including frames, scripts and schemas.

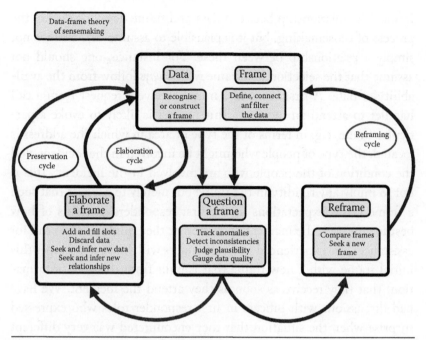

Figure 2.3 Sensemaking in the data-frame model.

may result from direct contradictions to the frame, the accumulation of discrepancies or the detection of subtle anomalies; questioning a frame may lead to elaboration (discovery of new data or relationships), frame preservation (explaining away anomalies) and the comparison of alternate frames or reframing (recovering discarded data and reinterpreting data). These tasks are summarised in Figure 2.3. Given the collection of tasks that are presented in Figure 2.3, it should not be surprising that sensemaking is a slippery concept to fully define. This slipperiness stems from the fact that each task contains its own actions that could be described as sensemaking. In broad terms, one could simply state that sensemaking is all of these tasks, with the expert sense maker placing emphasis on each task as the context changes.

In Figure 2.3, the top oval represents the basic sensemaking cycle of frame-defined data collection and data-based frame modification (Klein et al. 2007). Any of the tasks in Figure 2.3 can be a starting point for sensemaking, depending on the nature of the *surprise* or the perception of inadequacy of the existing frame that triggered it (Klein et al. 2007). Thus, the process of sensemaking can be viewed as being partly defined by the situation (data) and partly by the sense maker's frame. The relationship between data and frame develops during the process of sensemaking, but it is plausible to assume that there is not simply a relationship between these. For instance, one should not assume that the selection of a frame will *always* follow from the availability of data. A first responder might receive a request from a call handler to attend an incident. This request is likely to evoke a particular frame, e.g. in terms of the type of area in which the address is located, the type of people who might be involved in the incident, and the condition of the people who are involved in the incident. This is not to imply any rigidity in the frame but merely to suggest that such a frame raises expectations in the first responders in terms of how best to approach the incident and whether they might need to call for assistance. The experienced first responders will, however, temper this initial frame with a new frame that is built from the data (information) that they receive as soon as they attend the incident. We have had discussions with officers in first responder roles who expressed surprise when the situation that they encountered was very different from what they had initially anticipated.

2.4 Flexecution

If one aspect of sensemaking involves defining a situation in order to allow appropriate action to be taken, then it is useful to consider how *appropriate action* might be defined. We have noted in Section 1.1.1 that action could simply mean following the standard operating procedures (SOPs). In such instance, the role of sensemaking is to define the situation with sufficient clarity to ensure that a relevant SOP can be deployed. Under such circumstances, it is a moot point as to whether *sensemaking* is required – if the situation is sufficiently routine or familiar, and there is no gap in understanding, then one simply acts as usual. However, when the situation is unusual, then sensemaking is required. Furthermore, if the situation is sufficiently dynamic, then it is likely that the appropriateness of an action is going to change as the situation develops. This means that not only is it necessary to define an action, or a plan to follow, but one also needs to be able to adjust and modify the plan in response to changing situational demands.

Klein (2007) considered this ability to adjust and modify plans in terms of flexible execution (*flexecution*). He presents a view of plans that speak to the discussion in Chapter 3. For Klein (2007), plans are not simply 'roadmaps for solving problems' (p. 80) but are also frameworks that help determine what resources are needed, how these resources need to be coordinated, what risks might be associated with the use of these resources in this situation, where responsibility lies for managing the resources and risks, etc. Thus, plans serve as 'resources for replanning' (p. 80), i.e. they define the decision space in which activity can be monitored and evaluated. *'During flexecution, we're simultaneously trying to achieve goals and to discover, clarify and define them'* (p. 82). This implies a cyclical process (which sits comfortably with the data-frame model in Figure 2.3) that helps us to appreciate how the end of sensemaking need not simply be sense but can also be action. After all, the purpose of making sense of the situation (particularly in emergency response) is to be able to respond appropriately and confidently by deploying the best resources that are available in the best manner possible.

2.5 The Problem of Bias

The data-frame theory is studied in situations in which expert decision makers have relevant experience that enables them to rapidly make sense of dynamic situations (Klein 2011). As Leedom (2001) points out, the more difficult case is to make sense of novel problems and uncertain situations. Whenever one talks of decision making, it is almost inevitable that the subject of bias arises. For researchers in the NDM community, *bias* is a particularly problematic concept (Hoffman et al. 1995). Reports of bias typically come from experimental studies in controlled laboratory conditions in which participants demonstrate systematic deviation from a prescribed (rational) response to a given problem. The study of bias in real life (i.e. outside of these conditions) has produced quite-varied results, and it is not obvious that the forms of bias that are revealed in the laboratory are inevitable in other forms of decision making, or whether these arise from the artificial situations of the experiments or the lack of expertise of the participants (Schwartz and Griffin 1986). NDM is primarily concerned with decision making by experts, where their expertise is defined by the ability to sample information from an environment with which they have some familiarity. By *familiarity*, we do not mean that they know everything about the environment but rather that features and aspects of the environment will call to mind previous experience, in the form of frames, which can be used to make sense of what is happening. It is difficult to reconcile the notion of systematic bias in reasoning with this elicitation of frames. Of course, one might argue that, if a decision maker has a predisposition to overvalue certain features in the environment and consequently to consistently apply a frame, which leads to misinterpretation of the situation (and hence an inappropriate or wrong response), then this would constitute bias. We would agree – but would note that it is difficult to find evidence to support this.

There is also the problem that this use of the term bias is possibly confusing the definition of bias in reasoning and decision making with that from social psychology and attribution theory (a point that we will explore in Section 2.5.2). The third is that bias necessarily implies irrational decision making (on the assumption that the *correct* response would be arrived at through rational, logical decision making and that anything other than this is neither rational nor logical). We have a problem with this assumption in that it suggests that the aim of decision making is to arrive at an incontrovertibly correct solution to a problem. In the laboratory, this might be possible to define. In the real-life situations in which our sense makers find themselves, this is much more difficult to define. Indeed, the whole notion of a *ground truth* is often only something that can be arrived at *post hoc*. (And even then, there might be discrepancy over the salient details, whether these details were fully known to the individuals who are involved and whether there might have been conflicting demands that had a bearing on the manner in which decision making could be performed). For example, we were told of an incident concerning the rescue of workers from an offshore oil rig, involving three organisations. During debriefing, after the incident was successfully resolved, it became apparent that the commanding officer in each organisation believed that he was in charge of the whole operation rather than just the response of their organisation. In the next section, we will take each of these points and develop them in terms of what they have to say about sensemaking.

2.5.1 Frames and Biases

The suggestion that, from the data-frame model of sensemaking, people will seek to apply frames to fit data might sound like the concept of heuristics in decision making, particularly in terms of the work of Tversky and Kahneman (1974). Indeed, much of the work on heuristics in decision making uses the concept of *framing* to explain how the presentation of a set of information can influence the approach to problem-solving and decision making that a person can perform. For instance, Gigerenzer (1996) argues that frames are rational responses to uncertain information but are not applied in the formal logic that Tversky and Kahneman (1974) seem to assume. In this instance, a frame is the manner in which the data are presented (as opposed to the interpretation of these data by the person). This is an important distinction because, even though the word frame is used in both cases, for the heuristics decision-making literature, a frame is the external representation of data (i.e. it is how the data are presented to the person), whereas for the data-frame model, the frame is (usually) considered as the internal representation of data (i.e. it is how the person perceives the data). One implication of this distinction is that it is possible to present data in very different ways, which *could* lead to the person perceiving these data in different ways. Many of the experiments in the literature relating to bias in decision making involve presenting data in ways that can confuse or mislead the decision maker.

This notion of framing as external representation can be directly related to the design of the artefacts that people are using to access and record information. In other words, the artefact-driven sensemaking that we will discuss in Chapter 3 might have the undesirable consequence of presenting information in misleading ways or of requiring people to engage in tasks that distract them from their immediate focus of attention. Having noted that data can be framed in such a way as to mislead people, we also stress that this is *not* the same as people applying the wrong frame in their sensemaking. The expert sense makers, with a rich experience of their work and the type of problems that they typically encounter, are likely to be able to bring to bear a wide range of frames with which to interpret data. We would suggest that, even when the data are presented in ways that could mislead or confuse the inexperienced person, the expert sense maker is able to reinterpret the data by applying a plausible frame. In other words, expertise is often characterised by the ability to take the external representation that is presented to the person and evaluate this against an appropriate internal representation. One implication of this is that, whilst there might well be systemic bias in external representation, this need not directly result in bias in internal representation. Having noted the confusion in the use of the term frame between these approaches, it remains important to identify how and where bias might arise in sensemaking, if only to allow readers to recognise this potential and develop strategies to deal with this.

2.5.2 Sources of Bias in Sensemaking

Section 2.5.1 has proposed that a major source of bias in sensemaking stems from the external representation of data. In other words, the manner in which data are collected, collated and displayed to the sense maker might be sufficient to induce error in interpretation. Suppose that a geographical information system (GIS) displays a map of the local area, with the location of police cars marked on this map. From this, it ought to be straightforward to plan the movement of cars, perhaps to arrange for them to block off particular roads. Let us assume that the incident develops from one involving pursuit in vehicles to one involving pursuit on foot – with officers *not* having any form of Global Positioning System on them. In this case, whilst the position of the cars has been accurately captured, the movement of the police officers now becomes impossible to track, and the GIS becomes redundant. In this case, pursuit could be monitored through requesting regular updates or commentary from the officers as they run after the suspect. However, this is unlikely to provide accurate, real-time data. The point of this simple example is that the GIS provides excellent support for some types of sensemaking based on some types of data, but is not for all problems. Furthermore, it is possible that the incident commander, lacking real-time data, could still attempt to manage the incident through the information that is displayed on the map, perhaps by using the location of the cars and guessing the officers' current location. In this case, the inefficiencies created by using a partial view of the situation and attempting to fill in the gaps can lead to confusion and misdirection.

In the UK newspapers in recent months, there has been much debate about the crime reporting statistics. In part, this debate relates to the reporting of the Crime Survey for England and Wales in which a sample of 50,000 people are asked about their impressions of crime in their locality. This tends to show variation in the sense of risk and security that people feel, often couched in terms of *falling crime figures*. These data are not the same as those from the crime statistics that are recorded and reported by the Office for National Statistics, which collects data from police forces. In terms of the impact of crime reporting and the collating of crime statistics, there has been much debate as to what offences are recorded as *crime*. In terms of the sensemaking that is considered in this book, this raises a potential form of bias in that different police forces could conceivably record different activities in different ways. To take a simple example, one police force that we have worked with treats all incidents in which police officers are called to skirmishes outside school gates (usually involving schoolchildren fighting) as a *violent crime* (irrespective of whether injuries have been sustained). Whilst there might be sound reasons for such a classification (particularly in terms of calculating the costs and resource implications of these responses), it is clearly different from other forms of violence. In this case, the bias lies in the definition of the incident, and this, in turn, seems to arise less from planning and coordinating the response and more from the manner in which the response is resourced and costed. On the other hand, defining these incidents as a violent crime might result in high priority and rapid response to the schools, which, in turn, can prevent the incident from escalating and can also reduce the risk to the security of residents in that neighbourhood. In this case, the definition of the incident is determined not solely by the risk to individuals but also by the perception of crime that local residents might feel and the need to ensure a visible police presence. Indeed, it is not unusual for police forces to grade some fairly low-level activity as higher priority because this ensures increased visibility, which, in turn, is likely to reduce the occurrence of more serious crimes.

From this discussion, it should be clear that the definition of an incident type could be made using information outside that provided in the initial call. The policy and procedures of a given police force could result in incidents being classified in slightly different ways in different parts of the country. Taking a literal view of the notion of framing (from the literature on heuristics and bias), one could regard this as a form of bias. However, taking a more pragmatic view of the demands that are made on policing, an alternative explanation is that this represents the interpretation of data against a background on local imperatives.

More controversially, there are incidents in which police response can reflect social bias, against particular racial or ethnic groups or against women or children. In such cases, where it can be shown that response is not proportionate, and it is motivated by misjudgement of the situation or stereotyping the people who are involved, there is a case that needs to be answered. In their review of the shooting of Jean Charles de Menezes, an unarmed Brazilian living in the United Kingdom whom the police had mistakenly identified as a suicide bomber, Jenkins et al. (2011) demonstrate how the information available to the various decision makers and the actors involved in the incident lead to mistakes and misjudgements. This leads to a sequence of events in which sense was being built on a flawed foundation. In part, the actions that were being performed were compromised because a set of assumptions had not been directly challenged. The incident involved a number of decision makers, each with a partial view of the situation in which the available information was incomplete but was sufficient to support a course of action. Where the overall *system* failed in this instance was that opportunities to check and confirm information were either missing or not taken, and the result was the shooting of an unarmed, innocent man. One could interpret this in terms of various forms of decision bias in which the key decision makers exhibited confirmation bias. However, this would require the assumption that they had not sought further checks on the identity or behaviour of the *target*. This was not the case, so rather than showing bias in their reasoning, it makes more sense to consider the activity in terms of the pressure on those making the key decisions to protect the lives of civilians (following the recent bombings in London on July 7, 2005). In this respect, the actions were motivated by the belief that the target was a suicide bomber and that he was intending to detonate a bomb on an underground train.

2.6 Conclusions

This chapter has considered the manner in which individuals recognise the gap in their knowledge of a situation. For the expert, the defining features of the situation are interpreted in terms of prior experience (in the form of a frame), which is used to rapidly select an appropriate course of action. From this account, individual sensemaking is a process of rapidly finding a frame that matches the situation in order to define a response (as described in the RPD model). As the situation changes, the process of selecting and applying a frame becomes cyclical (as described by the data-frame model). The manner in which the situation is presented *could* influence the individuals' perception of such a gap. This means that the situation could either be presented through the lens of a given technology, or the response could be constrained by the artefact being used. In the next chapter, we will consider the positive and negative impact of artefacts on sensemaking.

3
Sensemaking with Artefacts

This chapter begins with the assumption that sensemaking involves the use of artefacts, which can serve as frames to assist in the structuring of data. Sometimes, the use of artefacts can be beneficial, by providing a structured means of gathering information (e.g. forms to be filled in), but this benefit can easily become a problem when the artefact is not well suited to the situation. The use of artefacts is presented as a form of distributed cognition, where *sense* emerges from the interactions between the person and the artefacts that are used.

3.1 Introduction

From the discussion of the data-frame approach in Chapter 2, it is possible to consider artefacts as a form of frame, at least in their role as external representation, which contrasts with the internal representation that is held by the sense maker. The question for this chapter is whether it is feasible to consider artefacts not merely as frames but also as agents in sensemaking. In other words, we could simply claim that *cognitive artefacts* (Norman 1991) provide benefits for individual cognition by allowing people to *offload* some information rather than having to remember it. However, the nature of cognitive artefacts also allows people to perform some cognitive actions *in the world* rather than *in the head* (Norman and Draper 1986). In this chapter, we present the concept of distributed cognition, which argues that cognition cannot be so neatly captured as an individual activity but rather as arising from the interactions between people and their cognitive artefacts. This raises a fundamental philosophical question about *when* or *where* cognition occurs and whether it is solely the function of individual cognition. A distributed cognition approach would be at odds with the individual cognition descriptions in Chapter 4 because it would argue that sensemaking is not solely a matter of the expertise of the individual sense maker but also emerges from the interaction between the person and the artefacts that they use. In this way, the argument relates to the discussion of *common ground* in Chapter 1 in that sense arises from the interaction between people and artefacts.

We begin this chapter with the observation that the environment and the artefacts it contains can shape the way in which cognition is performed (Zhang and Norman 1994; Hutchins 1995a,b; Scaife and Rogers 1996). From this, we contrast frames as schema or *internal representations*, with frames as artefacts or *external representations*. External representations can influence the strategies that individuals use to solve problems (Chase and Simon 1973; Larkin et al. 1980; Chi et al. 1981); for example, changing the layout of a puzzle can make it easier or harder to solve. This perspective also highlights the importance of interactivity in cognition, for example, players of Tetris and Scrabble can benefit from being allowed to manipulate and rearrange the playing pieces (Kirsh and Maglio 1994; Maglio et al. 1999). This points to the need to not only focus on the arrangement and design of the external representation but also consider the nature of the interaction between individuals and objects.

Not only does the external representation affect the way in which people interpret a problem, but it also affects the way in which they view the activity that they can perform. For example, Kirsh and Maglio (1994) distinguish between *epistemic* activity, which relates to understanding and solving a problem, and *pragmatic* activity, which relates to manipulating and rearranging the external representation. Their observation of Tetris players suggest that some of the activity was directed not so much towards the immediate solution to a problem as rearranging the pieces in order to allow the user to spot possible solutions. The suggestion is that epistemic activity would appear to be directed towards solving a specific problem, and pragmatic activity would appear to be directed towards arranging the space until a problem could be recognised. In previous work, we note how crime scene examiners will not only seek to recover evidence but also act on the scene in order to reveal evidence, e.g. through the use of aluminium powder to reveal prints on surfaces (Baber 2013).

In distributed cognition, cognitive processes are not viewed solely as internal mental processes but also are mediated by interactions with physical objects, which serve to support and transform cognitive activity (Flor and Hutchins 1991; Artman and Garbis 1998; Attfield and Blandford 2011). Thus, cognition moves from taking place in the head to in the world (Norman 1993) and becomes a property of the system rather than being contained within a single individual (Artman and Garbis 1998). This is possible because any unit – regardless of size – that is engaged in problem-solving can be defined as a cognitive entity (Perry 2003).

Any artefact in the environment can be designed or adapted to serve as an information-processing function. In order to reduce the load that is placed upon limited mental resources, individuals often make use of physical objects in the environment, for example, handwritten notes. These artefacts can serve as external memory cues during complex problem-solving, reducing the complexity of the task and the associated mental workload (Norman 1993). Unlike traditional descriptions of individual cognition, where representations of knowledge are held within the individual's mind, within a distributed cognitive system, artefacts themselves act as representations of task-relevant information, and the system arrives at its goal state by performing transformations upon these representations (Flor and Hutchins 1991; Perry 2013). The transformation of representations is achieved by combining, interpreting and representing information that is provided by both artefacts and individuals in the system – no single person controls this activity (Hutchins 1995a; Artman and Garbis 1998). It is the co-ordination of work and the flow of information between the components of the system that lead to the development of systems-level cognition (Artman and Garbis 1998; Perry 2003). Consequently, artefacts are viewed as representing the critical information within a work domain (Nemeth and Cook 2004). In order to study the nature of cognition at the systems level, researchers therefore focus on the role of observable external representations, the flow of information between components (artefacts and individuals), organisational structures and processes governing information exchange and the environment in which the system operates (Hutchins 1995a; Baber et al. 2006; Perry 2013). As a consequence, the distributed cognition approach may reveal cognitive processes that would not be found by research methods that examine individual-level processes, such as many studies of decision making, teamwork and situation awareness (Flor and Hutchins 1991).

3.2 Artefacts as External Representations

The important drivers behind the distribution of cognitive activity to elements external to the mind are the potential to reduce effort, improve efficiency and enhance the effectiveness of that activity (Kirsh 2013). Wright et al. (1996, 2000) distinguish between abstract information structures that act as resources and the artefacts that represent them. Kirsh (2013, p. 171) identifies seven ways that external representations can enhance cognitive functions, stating that they

1. Provide a structure that can serve as a shareable object of thought;
2. Create persistent (i.e. stable) referents;
3. Facilitate re-representations;
4. Are often a more natural representation of structure than mental representations;
5. Facilitate the computation of more explicit encoding of information;
6. Enable the construction of an arbitrarily complex structure; and
7. Lower the cost of controlling thought – they help co-ordinate thought.

Baber et al. (2006) investigated the role of artefacts in the process of crime scene investigation and drew a distinction between informal artefacts (which are used to make sense of a crime scene) and formal artefacts (which form part of the final report). In an investigation of collaborative sensemaking during emergency medicine, Reddy et al. (2007) differentiated between the *structured* (i.e. formal) articulation that is required by the computer support that they used and the use of low-tech artefacts (pen and paper, whiteboards) for unstructured (i.e. informal) articulation. The structured articulation was felt to overly constrain communications, with the result that during crises, staff would revert to low-tech alternatives, due to the greater ability to accommodate improvised working practices and information requirements (Reddy et al. 2007). Similarly, Khalilbeigi et al.'s (2010) study of technological support for command and control in large-scale disasters found that users retain an attachment to established work practices and low-tech artefacts, such as pen and paper.

Kirsh (2013) argues that for a representational system to perform a cognitive function, it must be *'sufficiently manipulable to be worked with quickly'* (p. 187). There is therefore a tension between the desires of organisations to use technology and the preference of many employees for the flexibility that is afforded by less formal alternatives. Whilst Reddy et al. (2007) allow for the possibility that artefacts facilitate sensemaking, they appear to view artefacts as merely transferring information between actors rather than as an integral component within sensemaking.

The use of artefacts to support sensemaking can be highly beneficial but can also create problems. Artefacts could not only distract the person from sensemaking (through the imposition of tasks that the artefacts require but are not central to the cognition of the sense maker) but also require a phrasing of the sense in a way that might be relevant to the artefact but might limit human ability to fully conceptualise it. This is like the problems of verbal protocol where people might phrase sense to make it easier to put into the form rather than in terms of their understanding. Once sense has *crystallised* (de Jaegher and di Paolo 2007) and can be recorded, then being able to store and send this information can be useful to an organisation, not only for current operations but also as a record of decision making for later review. This points to the perennial conflict between making sense and recording sense. Often, the role of artefacts seems to be designed with the purpose of creating a record of the sense (in terms of key information, key decisions, rationale for decisions, etc.) rather than as a way of supporting the activity of sensemaking. This raises the question, what properties do artefacts need to possess in order to enable or support sensemaking?

3.3 Artefacts as Part of a Cognitive System

> ... *the entities operating within the functional system are not viewed from the perspective of the individual, but as a collective. In the analysis, both people and artefacts are considered as representational components of the system, using the same theoretical language to describe their properties.* (Perry 2003, p. 206)

Hutchins' (1995b) study of navigation aboard a US navy vessel is seen as the definitive application of socially distributed cognition (Perry 2003). The task of navigation is *'an emergent process arising from the coordinated actions of the crew'* (Perry 2003, p. 197). These co-ordinated actions are socially, technically and temporally distributed and could not be reduced to the cognitive workings of individual crewmembers (Rogers and Ellis 1994; Hemmingsen 2013).

For Hutchins (1995b, p. 49), the artefact could be more than a display of information:

> *Having taken ship navigation as it is performed by a team on the bridge of a ship as the unit of cognitive analysis, I will attempt to apply the principal metaphor of cognitive science – cognition as computation – to the operation of this system. In doing so I do not make any special commitments to the nature of the computations that are going on inside individuals except to say that whatever happens there is part of a larger computation system. But I do believe that the computation observed in the activity of the larger system can be described in the way cognition has traditionally been described – that is, as computation realised through the creation, transformation and propagation of representational states.* (p. 49)

Hutchins (1995a) gives an example of the reformulation of a representation into a more transparent form in order to reduce the cognitive effort that is associated with a complex task – that of maintaining aircraft speed within safe parameters during landing. Figure 3.1 shows a drawing of a cockpit airspeed indicator instrument, taken from Hutchins (1995a); around the edge of the instrument are a number of speed *bugs* (pointers) that relate to the required air speed for various flight conditions (e.g. different wing configurations). The required speeds are calculated prior to commencing landing, based on instrument readings for the gross weight of the aircraft and a set of predetermined speed/weight calculations (Hutchins 1995a). During landing, the crew refer to the positions of the airspeed indicator needle that are relative to the bugs, in order to confirm that reductions in speed are co-ordinated with the appropriate wing configuration changes. In this way, the complex task of ensuring that an aircraft of a given weight maintains the appropriate wing configuration for its speed is translated into a simple visual check of the air speed indicator,

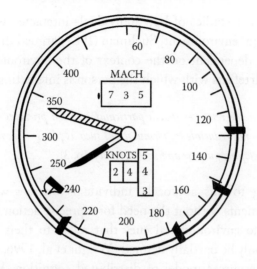

Figure 3.1 Speed bugs on an airspeed indicator.

which explicitly represents the relationship between the current state – needle – and the goal state – speed bug. The cognitive effort in calculating the various speed/wing configuration thresholds was carried out during a period of relatively light workload and encoded in a meaningful representation, which the aircrew could later draw on during a period of higher workload (Hutchins 1995a).

The notion that cognition/computation is *the creation, transformation and propagation of representational states* not only echoes in Section 3.1 but also implies that the representational states themselves will invite or encourage a particular response, as we will discuss in the next section.

3.4 Artefacts as Resources for Action

In addition to the wide range of ways that even simple artefacts can enhance the cognitive functions that are described in Section 3.3 (Vallée-Tourangeau and Cowley 2013), artefacts (and the abstract information structures they represent) are thought to be able to function as resources for action. The argument is that the design (in terms of physical appearance or functionality) of artefacts may act as prompts for agents to perform certain activities, without conscious reflection (Fields et al. 1996; Baber et al. 2006). This mirrors Klein et al.'s (2007) view of the role of frames in sensemaking.

Based on her studies of how individuals interacted with artefacts in their natural environment, Suchman (1987) argued that actions are situated, i.e. dependent on the context of the environment and the state of the artefacts with which the person is interacting:

> ... *actors use the resources that a particular occasion provides – including, but crucially not reducible to, formulations such as plans – to construct their action's developing purpose and intelligibility.* (p. 3)

According to this approach, individuals interact with artefacts and environments without the need for a precise action plan, instead responding to environmental cues that relate to their overall goal, which may only be partially defined (Wright et al. 1996, 1998, 2000).

In their resources model of distributed cognition, Wright et al. (1996, 1998, 2000) identify six different types of information structures that can be described independently of how they may be represented; these are (1) plans, (2) goals, (3) possibilities, (4) history, (5) action–effect relations and (6) states (Wright et al. 2000).* Each of these may be represented internally or externally in a number of ways (e.g. plans could take the form of memorized procedures or written instructions or might be incorporated into the design of an interface) thereby enabling artefacts to represent and convey the abstract information structures (Wright et al. 2000; Baber et al. 2006). Wright et al. (2000) view the use of representations as a cyclical process, whereby action is informed by the configuration of (both internal and external) resources; when an action is taken, this changes the configuration of resources, which prompts further consideration and action, and so on.

Various states and combinations of resources can inform action in a number of ways, which Wright et al. (2000) term *interaction strategies*. They identify the following four strategies:

1. *Plan following* (plan, history, state)
2. *Plan construction* (goal, possibilities, action–effects, state)
3. *Goal matching* (goal, possibilities, state)
4. *History-based elimination* (goal, possibilities, history)

---

* Though they acknowledge that additional types of structures may be identified.

This means that, from the point of distributed cognition, the variation in goal for different agents for the same goal has implications for how information is collected and shared. In distributed cognition, the resources for action need to be represented in a manner that can support collaboration, provide input to formal reporting and be adaptable to different contexts. These broad requirements can be taken further into the following specific requirements (Seagull et al. 2003):

- *Provide a common reference for communication* – This could either mean placing more emphasis on the final product, i.e. the report, and allowing agents access to it, or providing some means by which information can be shared by all parties as required.
- *Provide a 'communal memory'* – This could relate to supporting access to information rather than relying on recall.
- *Provide support for collaboration.*
- *Provide a mechanism for multiple manipulation of objects* – The products produced by different agents become the focus of different activities for other agents, which suggests a need for tracking of these activities on the objects (or version control of the objects themselves).
- *Support flexible content reconfiguration* – This could be interpreted as the need to allow different agents to modify information relating to the items as their analysis proceeds.

The resources model suggests that choices regarding strategy will affect the abstract information structures (and thereby resource) that are attended, whilst, at the same time, the resources are attended to inform the information strategy that is selected, i.e. artefacts function as *resources for action* by cueing particular behaviours and affording specific responses (Baber et al. 2006; Baber 2013). This description relates to Ramduny-Ellis et al.'s (2005, p. 76) distinction between the artefact as designed and the artefact as used, i.e. how people have 'appropriated, annotated and located artefacts in their work environment'. The *appropriated artefact* illustrates creative ways in which people use available materials for information recording or sharing. A common example might be the use of a stick to draw a map in the sand to explain a particular course of action. A more modern example might be the use of a motorcycle's fuel tank as an information store (Figure 3.2).

In Figure 3.2, a traffic officer has used a chinagraph pencil to convert his police motorbike petrol tank into an artefact. He has recorded information that is required for his intercept role during a large traffic operation, including geographic locations (motorway junctions), individuals and roles (names and call signs) and communications (radio talk groups). The notation on the fuel tank means that by glancing down, the officer is able to remind himself of these details whilst riding on the motorway, without having to pull over to check his pocket notebook or radio the controller. This practice seemed to be common amongst motorcycle traffic officers, but it is not formally taught.

Baber (2013) describes crime scene examination as a distributed cognition process in which the environment and the objects it contains become resources for action for experienced crime scene examiners, affording interpretations (such as cueing what evidence to recover)

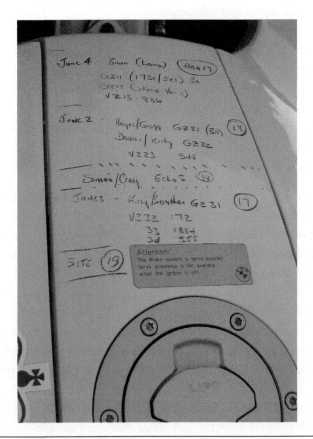

Figure 3.2 Notes made on a police motorbike fuel tank during a traffic operation.

that are not available to the uninitiated. The actions afforded by these artefacts will differ, depending on the training and experience of the agents within the system. Baber (2013) gives two alternative (*weak* and *strong*) views of the criminal investigation process as distributed cognition. Firstly, Baber (2013, p. 144) presents *the distribution of artefacts* in which objects '*function as vehicles for the storage or representation of information*' and are acted upon and altered by individuals within the system. Secondly, Baber (2013, p. 144) posits *the distribution of tasks* in which agents and artefacts are participating in a collective information processing activity, which is not necessarily centrally co-ordinated and that

> ... *accumulates information to a point at which its interpretation can be tested in Court ... The action of one individual will form the basis for actions of the next. In this manner, the criminal justice process is able to 'know' the collected evidence, even though it is unlikely that a single individual will have access to all of the information collected during the examination.*

Interestingly, whilst the crime scene examiners in this account may be seen as part of a collective activity, the geographic and temporal distribution of the agents within this process may preclude meaningful collaboration, thus differentiating this analysis from the concept of socially distributed cognition, which is discussed in Chapter 4.

3.5 The Problem with Sensemaking as *Representation Construction*

So far, we have considered the role of artefacts as external representations and as resources for action. Taking these notions could lead to the conclusion that sensemaking is all about creating and manipulating external representations. This is the position that Pirolli and Card (2005) describe in their model of sensemaking, which is summarised in Figure 3.3.

The rectangular boxes represent data flow; the circles represent the process flow. Pirolli and Card (2005) describe sensemaking activity as a series of iterative loops that form part of two sub-processes, firstly of foraging for information and then of developing a mental model to fit the information. According to this approach, in order for information to support expert assessment (and then the communication of that assessment), as it passes through the various stages in the process, it

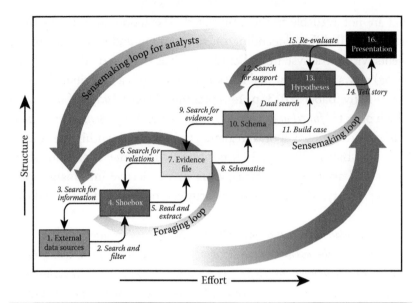

Figure 3.3 Notional model of the sensemaking loop.

is progressively transformed from raw information into the following intelligence products:

- *External data sources* contains all of the raw evidence that is presented to the analyst.
- The *shoebox* is a smaller subset of the total information and represents that which they deem to be relevant to the task.
- The *evidence file* contains small components that are extracted from items in the shoebox.
- *Schemas* contain information that has been reorganised or re-represented, in order that it can be used more easily to draw conclusions.
- *Hypotheses* are the initial representation of those conclusions (complete with supporting arguments).
- The *presentation* is the final intelligence product that will be shared.

Pirolli and Card (2005) define three broad activities that are involved in foraging for information: (1) exploring (increasing the scope of information that is included in the analysis), (2) enriching (narrowing the set of items that are collected) and (3) exploiting (extracting information and generating inferences). In terms of making sense of information, the key activities are thought to be (a) problem structuring (generating, exploring and managing hypotheses), (b) evidentiary reasoning (organising evidence to support or refute hypotheses) and (c) decision making. Schemata are seen as taking a central role within both the foraging and sensemaking loops (Pirolli and Card 2005). The process described in Figure 3.3 can be top down as well as bottom up, as new hypotheses can prompt the re-evaluation of schemas and source material and initiate the search for new information (Pirolli and Card 2005).

The notion that sensemaking is iterative is similar to the ethos of the data-frame model, explored in Chapter 2. However, there are some fundamental differences between these approaches. For the data-frame model, the development of sense continues as long as the sense maker has time and effort that are available to undertake this activity, and results in a *satisficed* (to use Simon's [1956] term to describe a decision that is sufficient for the task, information and demands at hand). For the proponents of the data-frame model, this approach represents the expertise of the sense makers that they have studied. For the Pirolli and Card model, sensemaking is far more about imposing order on data, in terms of developing a *product* that can be worked and shared. Not only does this imply a distinction between internal representation (in the data-frame model) and external representation (in the Pirolli and Card model), but it also shifts the nature of the activity from cognition and reasoning (in the data-frame model) to foraging and editing (in the Pirolli and Card model). Perhaps, the former focusses on the question of sense, and the latter prioritises *making*.

Attfield and Blandford (2011) describe the process of sensemaking in legal investigations as being broadly similar to Pirolli and Card (2005), again identifying two main processes – (1) data focussing (review and shortlisting of relevant information) and (2) issue focussing (identification and organisation of areas of enquiry). Sensemaking in legal investigation is also a bi-directional (i.e. top-down and bottom-up) activity, as theories are formulated and re-evaluated (Attfield and Blandford 2011). Both Pirolli and Card (2005) and Attfield and Blandford (2011) consider the transformation of external representations to be a key component of the sensemaking process: Pirolli and Card (2005) describe how intelligence analysts use artefacts, such as maps, databases and networks, to organise and understand information about people, organizations, tasks and time, in order to develop and communicate insights. Similarly, Attfield and Blandford (2011) give an account of legal teams creating and manipulating a range of artefacts in order to make sense of and communicate their findings during investigations. As such, Figure 3.3 provides a view of sensemaking as a process of the transformation of internal (10, 13) and external (1, 4, 7, 16) representations of information '... *from its raw state into a form where expertise can apply (such as for interpretation or taking action) and then out to another form suited for communication*' (Pirolli and Russell 2011, p. 3).

For Faisal et al. (2009), these external representations form the frames or schemas that guide and influence interactions with the data. Treating external representations as schemas moves beyond the purely in-the-head view of sensemaking and towards one in which sensemaking is a technologically mediated activity (Attfield and Blandford 2011). Consequently, sensemaking may be viewed as partly defined by the tools that are available.

This notion of sensemaking as product construction could be seen as a version of the distributed cognition that is outlined so far. However, whilst this recognises the importance of external representation, it seems to imply that the focus of the analyst is on making representations rather than making sense – the suggestion that the representation is the sense. We can see how this fits the need of the analyst to create product. But, this seems an extreme version of artefact-driven sensemaking in which the artefact becomes the constraint for the sensemaking.

3.6 Distributed Cognition and the Extended Mind

Rather than considering distributed cognition simply in terms of the ways in which people might make use of artefacts to assist in their cognitive activity, there is a more radical theoretical position that claims that the use of artefacts *is* cognition. In other words, using artefacts is not something that is separable from cognition, but the use of artefacts becomes an integral part of cognition. In order to appreciate the arguments that this position has aroused, it is useful to begin with one of most often cited examples of this theory: the case of Otto's notebook (Clark and Chalmers 1998).

Two characters, Otto and Inga, decide to visit an exhibition at the Museum of Modern Art in New York. Inga remembers that she knows the address of the museum and recalls that it is on 53rd Street and walks to the exhibition. Otto, on the other hand, has problems with his memory, so he writes everything in his notebook. He looks in his notebook for the Museum of Modern Art, finds the address and walks to the exhibition. Whilst this seems a simple comparison of recalling information from (biological) memory and recalling information from consulting an artefact, it has raised a number of issues in the philosophy of mind, and we believe that it offers some interesting points for considering *how* artefacts support sensemaking.

Rather than engaging with the entire debate on extended mind, we will concentrate on two issues. The first concerns the question of where *cognition* occurs in this example. This addresses the question of what Adams and Aizawa (2008) have termed the *mark of the cognitive*. The second concerns the question of how the different sources of information (memory or notebook) relate the beliefs of the characters. Fundamentally, Clark and Chalmers (1998) claim that the sentence in Otto's notebook 'The Museum of Modern Art is on 53rd Street' is *identical* with Otto's beliefs about the address of the museum.

3.6.1 *The Mark of the Cognitive*

Reading about Otto's notebook, one might be tempted to say that, of course, Otto is relying on some artefact to help his failing memory, and, of course, this artefact allows Otto to *offload* information into a repository that he can consult. The act of consulting the notebook (in terms of remembering or knowing that it contains relevant information, in terms of recognising the information that is stored in the notebook when he reads it, in terms of comprehending and acting on this information, etc.), all constitute *cognitive* acts and these are surely performed by Otto and not by the notebook. The reason for this assertion is that it seems obvious that cognition happens in the head (in this case, in Otto's head). This offers a lightweight explanation of the role of the notebook in *extending* Otto's mind in that the notebook holds information that is not held in Otto's memory. This is much the same as the way in which we might fill the address book of our smartphone with contact details and phone numbers, and then not attempt to remember any of these numbers because we know that they are all stored in the phone. Calling one of the people from the address book then simply involves recalling their name, searching the list of names and then making the call.

The reason why the case of Otto's notebook is an interesting thought experiment for the extended mind hypothesis is that there is a claim that the notebook is not simply an external artefact but is also essentially and integrally part of Otto's memory system, and because of this, the contents of the notebook constitute part of Otto's memory (and hence, part of Otto's beliefs). From this point, it is a small step to argue that, therefore, the notebook is an essential and integral part of Otto's cognitive system. For example, if Otto loses his notebook, he not only loses the ability to find specific information, but he also loses that part of his belief system that is constituted by the notebook. In other words, losing the notebook would not only mean that Otto could not find the address of the Museum of Modern Art (which the notion of notebook as information repository implies), but he would also lose knowledge of the existence of the Museum of Modern Art (because he had lost this part of his memory). What this means is that the notebook is not merely a passive store of information, but it is also coupled with the rest of Otto's cognitive system in such a way that the loss of the notebook would constitute damage to this cognitive system (in much the same way that other forms of damage to the cognitive system would cause problems).

This latter assertion suggests that the notebook has a *structural* role in cognition over and above the *process* role of being a store that can be consulted. For Adams and Aizawa (2010), this structural role constitutes a *coupling-constitution fallacy*, which means that just because an artefact is consulted in cognition, it does not follow that this artefact is, therefore, part of any cognitive process. In order to understand why they need to make such a statement, it is useful to consider what the contrary position might be, i.e. that the artefact is part of the cognitive process. This is illustrated by the following quote from Andy Clark:

> *The brain's role is crucial and special. But it is not the whole story. In fact, the true (fast and frugal!) power and beauty of the brain's role is that it acts as a medicating factor in a variety of complex and iterated processes which continually loop between brain, body and technological environment. And it is this larger system which solves the problem ...* (Clark 2001, p. 132)

What strikes us as interesting in this quotation is (a) that the notion that the *brain, body and technological environment* constitutes a *system* and (b) that the system engages in *complex and iterated processes*. In particular, this latter point accords with the data-frame model that we considered in Chapter 3. Where the extended mind concept departs from the data frame (and other individual sensemaking concepts) is in the manner in which cognition is performed. Adams and Aizawa (2010) claim that the extended mind hypothesis conflates information processing with cognition. This has been an issue with cognitive sciences since people began using the computer metaphor in the 1960s. It is perfectly possible to have information processing without cognition, e.g. think of the way in which a temperature sensor passes data to a microprocessor or the way in which a compact disc player reads the content of the disc – in both cases, information is being processed, but there is no cognition. For Adams and Aizawa (2010), the *mark of the cognitive* is defined in terms of specific processes that involve *non-derived content*, i.e. knowledge that is brought to bear by a cognizer. They use the example of a cane that is used by a blind person to aid in navigating and point out that the cane has no knowledge of its environment or the route to follow. Similarly, Otto's notebook does not have knowledge of the Museum of Modern Art. Whilst this might seem obvious, their critique misses what we believe is the crux of the extended mind argument, which is that meaning becomes constructed through interaction. This raises two points that are relevant to our discussion. The first is that the manner in which the interaction is performed becomes crucial in the way in which meaning is constructed. In other words, the way in which an artefact displays information and the way in which a person consults or responds to this artefact become significant in the construction of meaning. This is much the same argument as we rehearsed in Chapter 1 in the discussion of common ground. The point is not so much that the *meaning* exists in the form of some fixed content (in the person's head or in the display of information) but that there is an interaction that leads to the meaning becoming meaningful.

3.7 Conclusions

The artefacts we use to sense, process, collect, store and present data and information serve as external representations of the world with which we interact. For distributed cognition, these artefacts are not merely passive objects but also become part of the system that is engaged in making sense of the situation. At one level, this implies that missing or erroneous external representations can lead to problems in sensemaking. At another deeper level, this implies that the artefacts serve to shape the manner in which sensemaking is performed. A more extreme version of this notion is that the artefacts become active participants in the sensemaking process, and their role in not only to display information but also to suggest solutions or courses of action. The simple example of a form to be filled in illustrates this. The sequence in which questions are answered could be entirely dictated by the form's structure. For the expert, the form might be less important than following the conversation or detecting features of the situation, and then using the form to record the information. (The expert might even dispense with the form *per se* and seek other means of capturing and sharing the information.) The question of what is shared and how sharing occurs is considered in the next chapter.

4
COLLABORATIVE SENSEMAKING

If sensemaking relates to common ground, and *sense* is made through interactions, then it follows that collaboration can play a key role. In this chapter, we consider ways in which social, collaborative and organisational factors influence sensemaking. This chapter uses Weick's notion of collaborative search after meaning to develop the discussion of how sensemaking is performed as a social activity. Following this, we consider the nature of situation awareness and explain how the concept of distributed situation awareness provides a useful bridge between the artefact-driven sensemaking that is discussed in Chapter 3 and the collaborative sensemaking that is explored in this chapter.

4.1 Introduction

In contrast to the individual cognition or artefact-driven views, other approaches assert that sensemaking is a social process whereby a common understanding of the task and mental representations are distributed across the members of a group (Thompson and Fine 1999). This *socially shared cognition* interpretation takes the view that external media (artefacts) are only involved in social processes to the extent that they are co-opted by the group for communication purposes, to enable the sharing of perceptions, beliefs and intentions to create similarity across individuals' cognitions (Heylighen et al. 2004). Weick (1995) views sensemaking as firmly grounded in social activity, taking the organisation as the level of analysis. When viewed in terms of collaborative networks (i.e. communities of practice and exploration networks), this may explain how sensemaking can be conducted as a systems-level activity. Socially distributed cognition is primarily concerned with the properties of wider systems that emerge through the co-ordination and communication of human agents. In other words, socially distributed cognition '...*includes phenomena that emerge in social interactions as well as interactions between people and structure in their environments*' (Hollan et al. 2000, p. 177).

Information processing at this level therefore requires people to co-ordinate their activity in order to spread the cognitive load across the group (Perry 2003). This is achieved through communication, and language is also seen as a cognitive artefact (Perry 2003). The term *cultural heritage* is used within the socially distributed cognition tradition to refer to the way that the adoption of pre-existing artefacts, strategies, processes and procedures shapes activity within the workplace (Hutchins 1995b). By describing systems-level cognition in terms of the communication, and transformation of representations, it is thought possible to identify problems with the current process – such as information bottlenecks and breakdowns in communication – thus providing opportunities for improvement (Perry 2003).

The characteristics of socially distributed cognition systems can be identified from the literature (Rogers and Ellis 1994; Hutchins 1995b; Perry 2003, 2013) and can be summarised as follows:

- *Redundancy*: In the event that a single component fails, other media (artefacts or agents) prevent critical system failure.
- *Social organization*: Tasks must be organized such that they can be divided into components that can be performed by individuals, before being reintegrated again.
- *Adaptation*: People reorganize the environment within which information processing takes place, so social, cultural and historical elements become important components of the system.
- *Co-ordination*: Collaborative information processing involves information from several sources and different formats; this involves the combining and cross-referencing of different forms of representation and requires shared access to information.
- *Common representation*: For artefacts to be utilized by a distributed cognitive system, they must have a universally understood meaning, derived from common experience, training or context.

These characteristics raise questions for incident response command and control networks, for example, how are social organisation and co-ordination managed in incident response? Is there opportunity for adaptation within the formal and prescriptive processes? Is adaptation more prevalent during major incidents, where the responding organisations are dealing with novel situations? How is common representation defined and managed?

Landgren (2005a) makes the case for a common representation of perceived incident location in order to resolve delays resulting from information ambiguity. By providing richer information (i.e. from multiple sources) to responding crews, Landgren (2004) argues that collaborative sensemaking between dispatchers and responding units would be enabled. For this section, we concentrate on the social organisation and co-ordination aspects.

Landgren (2004, 2005a) described the effect of ambiguous information on incident response from a fire and rescue perspective. He describes how the incorrect framing of the incident delayed the resolution of the incident, particularly as the crews became committed to a *wrong* response. He suggested that the command and control, centralised at the command centre, conflicted with the need for the responding units to have a *preferential right of interpretation* of the situation. This led to a tension, as the responding crews have no direct access to information (relying on filtered, second-hand information) and were unable to validate or find inconsistencies, which raises questions over how key information should be captured and shared. In this respect, sensemaking could be closely allied with the concept of situation awareness. Before considering the link between sensemaking and situation awareness, we explore in a little more detail the ways in which sensemaking can be a collaborative activity.

4.2 Collaborative Search after Meaning

Weick's (1995) framework for sensemaking provides a clear framework for the notion of collaborative search after meaning.

4.2.1 Identity: Perception of the Environment Is Affected by the Perception of Self or Group

Sensemaking is not simply a matter of processing information in order to make a product but is also concerned with the ways in which people define their expertise and capabilities. One implication of this is that the manner in which a situation is described will relate to the terminology that the group uses. On the one hand, this reflects the common training and language of a given group, making communication easier within the group but which could lead to confusion if the information is passed on to other groups (as we discuss in terms of interoperability in Chapter 11). On the other hand, it helps to define group membership, i.e. people who use that terminology are *in-group*, which could influence assumptions about the credibility, trustworthiness or reliability of the information that is provided by other people (Cornelissen 2013). A second implication is that the group identity could influence the features of the situation that are regarded as salient. Just as we spoke in Chapter 3 about objects as resources for action (in that they provided opportunities for particular actions to be performed on or with them), situations can become resources for action in that the particular skill set or range of equipment of a given group can become important in how they decide to define a situation, which then has a bearing on their sensemaking. Equally significantly, if a group does not have particular capabilities or knowledge of particular equipment, their ability to make sense in a way that is relevant and useful to others might be compromised.

4.2.2 Retrospective: Sensemaking Is Concerned with Making Sense of Events That Have Already Happened

In terms of incident response, a clear problem is to make sense in a way that allows responders to plan an action and predict the consequences of their action. However, if sensemaking is retrospective (because it relies on the interpretation of feedback from the situation), then there will either be a lag in sensemaking (between the current sense and the current *situation*), or there will need to be an effort that is made in terms of developing and checking assumptions that will inform predictions.

4.2.3 Enactment: The Process of Making Sense Necessitates Active Involvement with the Environment and the Situation

Sensemaking requires people to change the situation in order to confirm that their sense is appropriate and in order to update their sense.

This corresponds to Landgren's (2004) distinction between *committed interpretation* and *committed action* and illustrates the need for sensemaking to be more than a purely conceptual exercise, particularly in terms of incident response.

4.2.4 Social: Making Sense Involves the Creation of Shared Meaning and Shared Experience That Guides Organizational Decision Making

The role of others in sensemaking becomes important as a means of guiding organisational response. As we noted in Chapter 1, some of the technology that people use in incident response or some of the products that they create might not have immediate benefit for their own sensemaking activity, but rather becomes important in subsequent activity in the organisation. These subsequent activities could be logically related to the incident response, e.g. in terms of providing situation reports to colleagues as an incident unfolds, but others could feel unrelated to the current operations, e.g. in terms of providing a log of activity that could be consulted during an after-action review or in subsequent investigation.

4.2.5 Ongoing: Sensemaking Is a Continuous Process That Starts before and Continues after an Event

Sensemaking is not a single event in which a single thing called sense in made. Rather, it involves the development and understanding of a given situation, for a specific purpose; in the case of incident response, the purpose is to ensure appropriate *control* of the situation (by which we mean the appropriate provision and co-ordination of resources). Not only is sensemaking ongoing in terms of its relation to a developing situation, but it is also ongoing in the more radical and richer meaning of continually emerging from the interactions between the sense makers. As de Jaegher and di Paolo (2007) have argued, sensemaking (at least in face-to-face dialogue) is enacted by the participants in their efforts to make sense (and through this effort to make meaning) during their conversations. Not only does this relate to the dynamics of the conversation, e.g. turn-taking, or to the development of common ground (see Chapter 1), but it also underlines the manner in which sensemaking is a participatory activity: '...*the coordination of intentional activity in interaction…*' (p. 497). Recall the second telephone call in Chapter 1 in which the caller seems hesitant or unable to provide useful information to the call handler. From the participatory sensemaking perspective, this is understandable as a process of developing a shared view of the nature of the conversation, i.e. establishing what the caller knows and how he or she should express this information in a way that is most useful for the call handler. What is important to note here is that it is not just the content (semantics) of the information that becomes important but also the manner in which this is expressed (pragmatics).

4.2.6 Extracted Cues: Information Is Provided by Interactions with the Environment; This Prompts Further Data Collection

The situation (or environment) contains cues that are discovered and responded to by people engaging in the situation. In collaborative sensemaking, the challenge is to ensure that all people engaging in the situation are sufficiently aware of the use to which their information could be put, e.g. in terms of providing clear and ambiguous information. This awareness can also, of course, impact on the nature of the information that is provided, e.g. people speaking to the police might be reluctant to provide information that they believe might incriminate themselves or might be scared of providing information that could elicit reprisal from others.

4.2.7 Plausible Rather than True: Sensemaking Generates a Coherent, Reasonable and Memorable Understanding of an Event That Guides Action, Rather than Attempting Accuracy

In Section 4.2.1, we considered the problems that are associated with a view of sensemaking as *building product*, which could then become the basis for a response. The suggestion that it is less important to have *true* sense than to have a plausible basis for proposing a response might feel counter-intuitive, particularly in the context of police response in which it is important to ensure adequate, appropriate, proportionate response to an incident. We suggest that this point highlights a distinction between what might be termed *informal sense*, which describes the evolving understanding of the situation and the formal reporting, and logs, which capture the response that is made and the basis for the response. Recall the second example in Chapter 1 in which a caller gives a vague and incomplete initial description of an event. If the call handler wrote this information verbatim in the log, then either the log would be confusing or there would be a need to keep editing it until a clear version was made. Of course, the issue of *translation* could become problematic if it was shown that the log contained erroneous or incomplete information. Consequently, it is common practice to maintain an audio recording of the call as a secondary source of information (although the recording is only stored beyond the duration of the incident if there is likely to be further investigation and although the recording is deleted once the incident has been resolved).

As with the data-frame model of Klein et al. (2007), Weick (1988, 1995) holds that the meaning and significance of elements within a situation may be open to different interpretations. For Weick (1988, 1995) and Gioia et al. (1994) sense makers are collaboratively involved in the interpretation of events through three aspects:

> *First, sensemaking occurs when a flow of organizational circumstances is turned into words and salient categories. Second, organizing itself is embodied in written and spoken texts. Third, reading, writing, conversing, and editing are crucial actions that serve as the media through which the invisible hand of institutions shapes conduct….* (Gioia et al. 2004, p. 365)

Sharing understanding involves making explicit, public, relevant, ordered and clear that which is otherwise tacit, private, complex, haphazard and historical. This is achieved through interactive dialogue, drawing on language '*…in order to formulate and exchange through talk…symbolically encoded representations of these circumstances*' (Taylor and Van Every 1999, p. 58). Such dialogue requires a shared language and some degree of common ground (Johannesen 2008), which we have previously considered in Chapter 1. Mutual knowledge is thought to take the form of shared context, such as domain, team, historical or artefact/environmental (Johannesen 2008). Johannesen (2008) provides a detailed description of how operating theatre staff work to maintain common ground during operations, drawing on a range of explicit and implicit communication strategies. This allows them to make joint assessments, collaboratively solve problems and quickly detect and recover from inappropriate actions (Johannesen 2008). However, Johannesen (2008) notes that when team members have common ground, '...*less needs to be said because information can be communicated relative to what is already mutually known*' (p. 198). Additionally, mutual knowledge means that verbal communications are characterised by their *compactness*, i.e. phrases, words and gestures carry meaning that is not generally accessible by lay persons without additional explanation (Johannesen 2008).

Similarly, Heath and Luff (2000) describe how when London Underground control room staff collaborate to resolve problems, they rarely provide explicit information to one another; instead, they monitor and respond to one another's actions, through reciprocal monitoring of activity and the use of shared artefacts. This process is enabled by their awareness and maintenance of a body of practice (procedures and conventions) relating to co-ordinated action, which '...*informs the production, recognition, and coordination of routine conduct within the line control room*' (Heath and Luff 2000, p. 102). This notion of mutual knowledge or a body of practice would seem to apply to sensemaking during the single-agency incident response activity, where team membership is relatively stable and members share common training and experience and use technical jargon to communicate. In contrast, Umapathy (2010) argues that collaborative sensemaking takes place when '...*a group of people with diverse backgrounds engage in the process of making sense of information rich, complex and dynamic situations*' (p. 1).

4.3 The Problem of Situation Awareness

As the process by which the range of possible interpretations of a situation is reduced to a single, most plausible explanation, sensemaking would appear to be a precursor to situation assessment. A commonly cited definition of situation awareness (SA) defines it in terms of

> ...*the perception of the elements in the environment within a volume of time and space, the comprehension of their meaning and the projection of their status in the near future.* (Endsley 1995, p. 36)

In terms of the preceding discussion of sensemaking, one can see how this could describe an aspect of the processes that are captured in the data-frame model (see Chapter 2), possibly in terms of the manner in which a frame is constructed for the *elements in the environment* that the sense maker is selecting and sampling. However, this definition of SA appears to offer a somewhat passive aspect of information processing (Leedom 2001; Turner 2007) in which the person receives (rather than actively seeks) information and *projects* (cognitively), rather than acts upon, the situation.

SA emerged as a key theme during Blandford and Wong's (2004) investigation of emergency medical dispatch – a complex collaborative process that involves discrete computer-supported and paper-based phases of activity. Senior staff described their SA as a *picture in the head*; this awareness enabled them to determine the type of units to allocate to an incident and to estimate the locations of resources (Blandford and Wong 2004). In this respect, the allocators could be seen as using some notion of a schema or *frame* to hold current SA. Allocators employed a number of strategies to maintain and refresh this awareness of the situation, drawing on colleagues and physical artefacts as resources in particular (rather than relying solely on their ability to memorise the situation). Furthermore, allocators report actively attending to other activities within the control room, which they refer to as *control ears*, in order to notice early cues (such as vehicle and job status changes) that they use to plan activities ahead of formal notifications. Additionally, allocators frequently refer to the physical location of tickets and their position in relation to one another to maintain awareness of allocated (in the allocator's box) and unallocated (laid out on the desk) incidents (Blandford and Wong 2004).

4.3.1 Distributed SA

One suggestion is that SA is analogous to the schema or frame that the sense maker is using to interpret a situation (Klein et al. 2006a). Rather than simply holding this SA in memory, it is clear from the Blandford and Wong (2004) study that people complement their ability to attend to and remember relevant data in a variety of ways, and that these ways appear to involve various forms of distributed cognition. If SA is, in a sense, the schema that an individual uses to inform sensemaking, then it becomes difficult to imagine how this could be shared by other people or how the concept of *shared SA* is possible. One approach might be to reflect SA in a product, perhaps a shared map (with pins indicating the location of particular actors) or a common operating picture. This takes the notion of product-driven sensemaking and develops this into a strategy for developing products to support activity. This is considered in detail in Chapter 13. For this discussion, however, we will point out that, as we have already argued, the product does not contain sense as such but can only serve as the basis for activity through which sense is made. In other words, product-driven sensemaking *must* involve interaction between people and product before sense is produced (even if this means that the people are simply reviewing or recalling what they have already interpreted).

This need for interaction to make sense has been considered in the collaborative sensemaking discussion and leads to the second approach to the ways in which SA is relevant to joint activity. Endsley and Jones (1997) defined team situation awareness as 'the degree to which every team member possesses the SA required for his/her job' (Endsley 1995, p. 39). The implication is, as Gorman et al. (2006) note, that the team SA arises from some sort of 'averaging, summing, or assessing the degree of overlap across individual mental states' (Gorman et al. 2006, p. 1315). This is problematic as, for example, Bolstad and Endsley (2003) sought to measure shared SA (in terms of overlap between knowledge that is held by individuals in a team) and found that 'only a few pieces of information were being shared consistently within cells'.

The concept of shared SA rests on a similar misunderstanding that characterises misuse of the term *common ground*. In other words, there is a presumption that *shared* or *common* means identical, and that the objective of creating shared SA or common ground is to ensure that all parties hold the same schema (knowledge structure) that they can access and deploy in the same manner. A variation on this presumption is that a common operating picture presents a representation that all people will interpret in the same way, extracting the same information and the same conclusions. These presumptions all rest on the fallacy that (a) information is equivalent however it is accessed and (b) that all people respond to a piece of information in the same way. An obvious test of this fallacy is to consider what happens when responders from different agencies join together in a multi-agency response. Even when people might have similar experience and training and be operating in the same space (say, a crew in an aircraft), there remains the potential for misunderstandings to occur.

Sharing, therefore, could look like an overlapping of knowledge with some people having the same schema or frames in their view of the situation (Figure 4.1). Such a view is implied by the notion that what is shared is SA (Endsley and Robertson 2000) or knowledge (Eccles and Tenenbaum 2004). However, as we have already noted, we find the idea that schema can be shared problematic – even if this means that the schema could overlap. The reason for this is that it does not make sense to view *schema* as anything other than highly idiosyncratic knowledge structures, and the notion of *overlapping* implies that the structures, somehow, exist independently of the people who build them. An alternative explanation of the Bolstad and Endsley (2003, p. 369) finding might be that SA does *not* 'exist only in the cognition of the human mind', as they claim, but that it arises from the interrelationship between different entities in a system. This perspective, team SA, should not be considered as the aggregation of individual SA but as the distribution of information through a system. Thus, Walker et al. (2009) show how presenting operators with richer and more detailed information can lead to a *reduction* in SA, not because the information is lacking but rather because the individuals do not need to process information that is being displayed in their environment. In other words, there is a distinction between knowing that the information exists on a specific display and seeking to process that

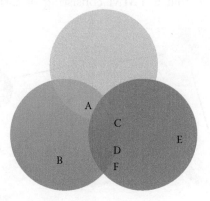

Figure 4.1 Representing shared (overlapping) SA.

information as *cognition in the human mind*. This echoes the notion of *transactive memory* of Wenger et al. (2002).

Rather than using *sharing* as a way of describing SA, we prefer to use the term distributed and have been developing a theory of distributed situation awareness through work with Stanton and colleagues (Stanton et al. 2006; Stewart et al. 2008). Figure 4.2 illustrates this concept: three actors have their own representations of information {A...F}, and these representations are logically related (perhaps in terms of an ontology). However, the connection between information need not imply either that the actors have the same information or the same overall understanding of this space. For example, the *red* actor does not have information A or B, or the *blue* actor does not have D, E or F.

This picks up on what Weick (1995) referred to as sensemaking being *plausible* rather than *true*; we do not mean that each individual is working with untrue information so much that, collectively, people have some variation in what they know, understand and interpret. This further implies that the frames that individuals apply to the situation can result in sense that might be distinct or that might, to a greater or lesser extent, overlap. To complicate matters, as a situation becomes more uncertain, so is the likelihood that individuals will have the same view of the situation, and hence being in a position to create shared knowledge becomes much less likely (Bourbousson et al. 2011).

In their review of team mental models (TMMs), Mohammed et al. (2010) propose that a TMM consisting of distributed *taskwork*

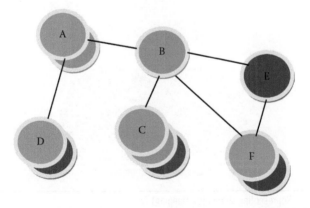

Figure 4.2 Representing distributed SA.

knowledge and overlapping *teamwork knowledge* could lead to improved performance. The suggestion is that there is a distinction between the information content, or the *semantic* aspects of SA, and the *pragmatic* aspects of SA, which relate to information sources, communication patterns, etc. (Duffy and Baber 2013). In a similar vein, Gorman et al. (2006) argue that it is essential to understand team co-ordination (in response to changes in the situation) and develop a co-ordinated awareness of situations by teams (CAST), to highlight how team dynamics change in response to the situation and how such dynamics seem to reflect the teams' trajectory towards a *common and valued goal* (Salas et al. 1992).

4.3.2 The Role of Artefacts in Distributed SA

In part, the concept of distributed SA reflects the notion of distributed cognition (particularly in the manner that it is used by Hutchins [1995a,b]) in which actors and objects are seen as discrete (i.e. unconnected) repositories of information. This means that there is no need to make any assumptions about the nature of the information (e.g. in terms of whether or not a particular piece of information is represented in the same way across two or more of the actors). It also means that the *meaning* of the information arises from the interaction between actors and between actors and objects. As Figure 2.3 illustrates, the connection between schema/frames, rather than being seen as a (passive) overlap of knowledge structures, becomes the (active) interaction between people through which knowledge is modified. This has the added benefit of allowing us to skirt around the vexed question of how such knowledge is stored, i.e. we tend to consider *knowledge* as something that is produced during and for the interaction rather than something that is a thing that can be stored. If there is a need to store this thing, then sense makers are likely to turn to the products that they have available to them in order to make a record (although we do not ignore the fact that the product could conceivably be the frame that the sense maker is using). This might lead to modification in the information that is held by an object (e.g. in terms of updating the value that it displays) or the frame being deployed by an actor. However, it does not mean that these modifications need to be aligned or, indeed, correct.

4.4 Sensemaking as System Activity

Whilst the discussion so far has been focussed on the description of sensemaking as an organisational and business activity, and has drawn primarily from sociological and organisational behaviour literature, an alternative perspective on sensemaking can be found in the application of systems dynamics models to this activity. For example, Johnston (2005) developed a systems dynamics model (Figure 4.3), following a fascinating ethnographic study of intelligence analysts' work activity. The intention behind this model was to reflect the

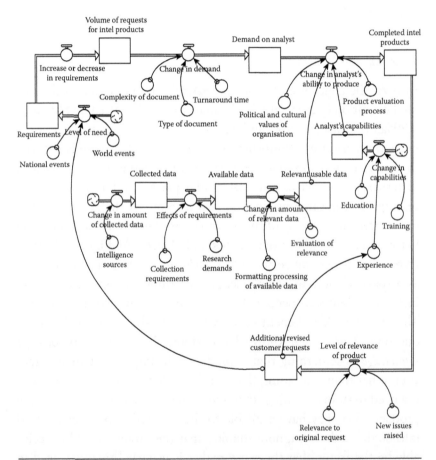

Figure 4.3 Dynamic systems model of intelligence analysis processes. (From Johnston, R. 2005. *Analytic Cutlure in the U.S. Intelligence community: An ethnographic study*, Washington, DC: Center for the Study of Intelligence. Available at https://www.cia.gov/library/center-for-the-study-of-intelligence/csi-publications/books-and-monographs/analytic-culture-in-the-u-s-intelligence-community/analytic_culture_report.pdf.)

constraints, demands and opportunities that impact upon successful intelligence analysis. We have found this analysis a very interesting and useful way to think about the intelligence cycle. In some sense, the model captures the cyclical nature of sensemaking of the dataframe model, with the various feedback loops and interconnections between the different aspects of the process.

Systems dynamics models describe processes in terms of stocks (which are containers for material) and flows (which transport material between stocks, based on some control parameters). In Johnston's model, *demand* originates from a request from policy makers or intelligence consumers, in response to some need, say, to understand the likelihood of a threat, or the political climate in a given regime, or the identification of suspects in a terrorist plot. In any intelligence analysis process, there are likely to be several such requests (although each request might have different demands that are associated with it, in terms of the workload that is required, the accessibility of materials, etc.). The requests modify the demand on the analyst (in terms of the type of report, the complexity of report, the changes that are required or the time to produce the report, etc.). In response to requests, the process engages in the *production* of reports and other outputs (product). The production activity will create a demand on the analyst, which draws on the stock of analyst capabilities. The resulting report is then evaluated, which depends on the nature and methods of this analysis process. The readers are advised to consult Johnston's (2005) book for a more detailed explanation of the model and the results that it produces.

Of interest to our discussions of sensemaking (which we are, in this section, viewing as analogous to some of the processes that are involved in intelligence analysis) is the manner in which the dependencies and interrelationships between the demands made on analysts, the knowledge, skills and abilities of analysts and the organisational culture of the organisation in which they work, create a set of constraints and opportunities that continually affect each other. This links back to the notion of different forms of *rules* that Manning (1988) discussed, both formal and informal, which influence the space in which sense is made. It also relates to the manner of the elements of Weick's collaborative search after meaning.

4.5 Conclusions

In this chapter, sensemaking has been considered in terms of arising from collaboration between individuals. Rather than this resulting in a combining of the sense that each individual brings, the collaboration leads to interactions that create new forms of sense. A key issue in collaborative sensemaking is the manner in which this new form of sense is captured and shared. For Weick (1995), this takes the form of a story that members of the sensemaking in-group are able to tell and elaborate and that contains the salient information. This implies a loosely structured account of the situation that people share verbally. From informal observations of, and participation in, the *huddles* that often occur during emergency response on the ground (see Chapter 11), it is often the case that informal information sharing and the development of the *story of the event* play an important role in allowing respondents to create a shared understanding of the situation. Many of the case studies that have been reported since Weick's (1995) pioneering work draw on analyses of discussions and meetings in which groups make sense of the problems that they face. Thus, there seems a strong case to be made for the proposal that collaborative sensemaking follows the elements that are outlined in this paper. However, as we noted in Chapter 1, the idea that there is an *informal* sense that can be used to describe and define a situation only covers part of the processes that sensemaking involves. For many situations (and this is often critical in emergency response), there is a parallel requirement to produce a *formal* statement of the response, and this requires a description of the situation in terms that can be used to justify the use of resources. There remains, we feel, a tension between the collaborative search after meaning, which seems to arise spontaneously when groups of people engage in sensemaking, and the formal requirements for reporting that organisational structures place upon them. In order to consider this tension further, the following chapters present examples of the organisational structure and responses that are involved in dealing with emergencies and incidents.

5

COMMAND AND CONTROL IN THE UK EMERGENCY SERVICES*

In this chapter, we present an overview of the context in which the studies reported in later chapters were conducted. In emergency response, the purpose of the command and control (C2) system is to detect and make sense of the unfolding incidents in order to put in place the appropriate responses. During routine emergencies, these agents respond in a co-ordinated manner using established procedures, with minimal command oversight. In contrast, major incidents occur infrequently and require proactive command activity at all levels. Major incidents may involve several emergency services who must collaborate in order to deal with a unique and uncertain set of circumstances.

5.1 Introduction

Incidents featuring ambiguous or unexpected events make people's sensemaking efforts *visible* (Landgren 2005a), which means that the emergency response domain should provide fertile territory for the study of the phenomena. This chapter sets the context for the activity that is studied in this book. Whilst the observations were conducted in UK police settings, this chapter explores these in terms of the concept of C2. Ultimately, we propose that different C2 structures might support different forms of sensemaking activity.

---

* Information in this section is based on observations and interviews with personnel from Warwickshire, West Midlands and Gloucestershire police forces, as well as fire and rescue service incident commanders.

During incident response, units are dispatched to the scene of an incident whilst the call handler is still gathering and interpreting details; officers attending the scene will first act to take control of the situation and only then begin the process of establishing what has taken place. This description raises the question, where in the incident response C2 system does sensemaking take place? The design of emergency response networks seem to imply that sensemaking is a centralised function, as information is channelled to central agents (i.e. controllers), who then allocate the resources that provide the incident responses. However, incident response is a collaborative process involving a wider set of individuals, both suggesting that sensemaking takes place across the network and that it involves collaboration between agents. (This becomes even more pronounced in major incidents that involve several emergency services – each with their own C2 systems.)

5.2 Emergency Service Operations in the United Kingdom

In the United Kingdom, the emergency services operate three levels of incident response: (1) strategic, (2) tactical and (3) operational; these are commonly referred to as gold, silver and bronze (Figure 5.1). In the main, most responses to incidents are handled at the

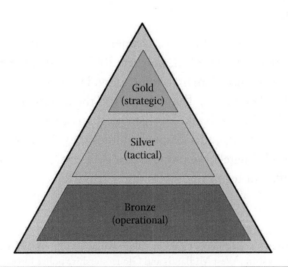

Figure 5.1 Emergency services major incident command structure.

bronze (*operational*) command level. Silver (*tactical*) command monitors activity at the force-wide level and only becomes involved if the complexity or severity of the incident requires the planning, co-ordination and decision-making activities that are associated with this higher command level (HM Inspectorate of Constabulary 1999). Finally, gold (*strategic*) command is only initiated when an incident has been, or is expected to be, declared a major incident. Major incident procedures emphasise liaison between levels of the command hierarchy and across the emergency services from the start of an incident,* so by the time that gold control is up and running, the three services should already have a response well underway at the scene.

Although there are a number of *Category 1* response agencies (the main agencies that are involved in the response to emergency situations; HM Government 2005a) the initial response to the majority of incidents in the United Kingdom is provided by the police, fire and rescue and ambulance services. Within each of these services, a number of separate organisations are responsible for the provision of response cover within specific geographic regions; in 2014, in England and Wales, there were 43 regional police forces, 50 fire and rescue services and 13 ambulance trusts. These are independent organisations, which means that there is no overarching national control authority for each service or for combinations of service (although major incidents, as we shall see, involve the creation of overarching control, and there are various groupings that can be formed on an as-needed basis). This means that, even within the United Kingdom, there is unlikely to be a single model of how an emergency organisation operates, with each organisation operating within its own culture and history, its own geographical region and its own financial constraints. This means that the observations reported in this book should be interpreted as instances of response in a specific organisation at specific points in time rather than necessarily as the *correct* or the *UK* way of doing business.

---

* LESLP. 2007. *Major Incident Procedure Manual*, seventh edition. Norwich: The Stationery Office. Retrieved May 2012 from http://www.met.police.uk/leslp/docs/Major_incident_procedure_manual_7th_ed.pdf.

Major incidents occur only very rarely in relation to the total number of incidents that are dealt with by the emergency services. A major incident is defined as a situation that requires special arrangements to be made by one or more of the emergency services, the National Health Service or the local authority in order to provide an adequate response (Cabinet Office Civil Contingencies Secretariat 2003).

A major incident may require the following (Cabinet Office Civil Contingencies Secretariat 2003):

- The initial treatment, rescue and transportation of a large number of casualties
- The involvement, either directly or indirectly, of large numbers of people
- The handling of a large number of enquiries that are likely to be generated both from the public and the news media, usually to the police
- The need for the large-scale combined resources of two or more of the emergency services
- The mobilisation and organisation of the emergency services and supporting organisations (e.g. local authority), to cater to the threat of death, serious injury or homelessness to a large number of people

By declaring that a situation is a major incident, the emergency services are able to allocate additional resources to the response. They can cancel non-emergency activities, can call up off-duty personnel and may draw on neighbouring forces to provide support for the duration of the incident.

Major incidents can require large numbers of resources from the three emergency services and other agencies and often involve working across large or hazardous sites. In order to enable a co-ordinated response with all personnel working effectively towards the agreed upon objectives, the emergency services implement the gold, silver and bronze command structure. Gold, silver and bronze are role-based designations and do not automatically equate to levels of seniority. During a response to a sudden, unanticipated major incident, this command structure will often be constructed from the bottom up, as any emergency service member is able to initiate a major incident response.

The process of establishing the major incident command structure is broadly similar for all three services. In the case of the police, the control room duty inspector would contact the duty assistant chief constable (ACC) to inform them of the situation, gold control is opened and the duty ACC assumes the overall responsibility for the incident (gold command). Meanwhile, another inspector would proceed to the incident site to act as the bronze commander. Major incidents can escalate rapidly, necessitating that the emergency services continue to respond whilst the command structure is being put into place. Until the designated bronze commander reaches the scene, the most senior attending officer is in charge. Once on scene, the bronze commander will assess the situation and form a plan of action to deal with the incident, including requesting appropriate additional resources.*

Within gold control, the senior representatives of all three emergency services (plus other agencies as required) will form the strategic coordinating group (SCG); they will discuss and agree on the high-level approach to the incident to ensure that there is a coherent response across all of the agencies that are involved. The police and other emergency services have contingency plans for different types of major incidents, which set out the broad structure and identify the initial priorities. However, each major incident takes place within a unique context, so the role of the SCG is to formulate the strategic response to the incident. This strategic intent is then translated into tactical plans via silver commanders and operational decisions on the ground via bronze commanders (Figure 5.2). This application of the principle of subsidiarity is similar to the military notion of mission command. The command structure also provides feedback as the situation changes or additional resources are required. An incident may be divided into different locations or types of activity, so several bronze commanders for each service may be present at the same site; this is reflected in Figure 5.1 by the widening of the triangle at lower levels.

---

* LESLP. 2007. *Major Incident Procedure Manual*, seventh edition. TSO, Norwich. Retrieved May 2012 from http://www.met.police.uk/leslp/docs/Major_incident _procedure_manual_7th_ed.pdf.

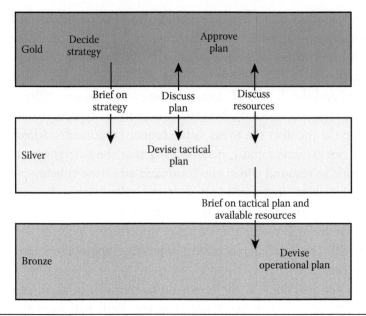

Figure 5.2 From strategic intent to operational decision making.

5.3 The Concept of C2

The term *command and control* originates in the military domain and, at its broadest, simply refers to the running of an organisation. The North Atlantic Treaty Organization (2010)* suggests that *command* encompasses the authority to direct forces. *Control* – the authority exercised by a commander – is therefore seen as a subset of command and may be delegated to others to perform. The difference between command and control is that command includes the authority and (by implication) ability to form new intentions (i.e. goals), whereas control involves delegated actions to achieve those intentions.

Alberts and Hayes (2003, p. 98) propose four minimum essential capabilities for C2:

1. The ability to make sense of the situation
2. The ability to work in a coalition environment
3. Possession of the appropriate means to respond
4. The ability to orchestrate the means to respond in a timely manner

---

* NATO. 2010. NATO Glossary of terms and definitions. AAP-6. Available at http://www.nato.int/docu/stanag/aap006/aap-6-2010.pdf.

Alberts and Hayes (2003) indicate that sensemaking is a prerequisite for any command system. As we have argued in Chapter 4, sensemaking is different from situation awareness (which is often used to define the manner in which a C2 system determines the nature of the environment in which it is operating). The second characteristic relates to the questions surrounding *interoperability* (which we consider in Chapter 11). The third and fourth characteristics relate to the manner in which resources are mobilised and deployed, which relates to the topic of control. In this book, we are less interested in the manner in which such control tasks are performed. The readers are advised to consult Faggiano et al. (2011) for an excellent guide on the management of incident response, which, in particular, covers the mobilisation and deployment of resources. In Sections 5.3.1 through 5.3.4, we consider the frameworks and models of C2 that are applicable to UK emergency services.

5.3.1 Police Incident Response C2 Organisation

In the United Kingdom, each regional emergency service organisation operates its own C2 network, which handles 999 calls and co-ordinates their response to the incident. These networks can be highly complex, with a large number of control rooms and different communications media in use. Figure 5.3 provides an overview of the command centres and main lines of communication that are used by West Midlands Police (WMP) during responses to incidents. (West Midlands Fire and Ambulance Services each operate separate C2 systems.) The WMP C2 structure is centred on the Force Communications Centre (FCC), which handles all incoming 999 calls, controls force-wide assets and performs the silver command role when required. The C2 network also features smaller local control rooms in each of the 21 operational command units (OCUs).* Communication across this C2 network is achieved via the electronic incident management system (IMS – used within and between control rooms), radio (control room to officers and officer to officer), and telephone (used with external organisations: fire and rescue, ambulance, etc.).

---

* Following the completion of data collection, the force was restructured into 10 local policing units.

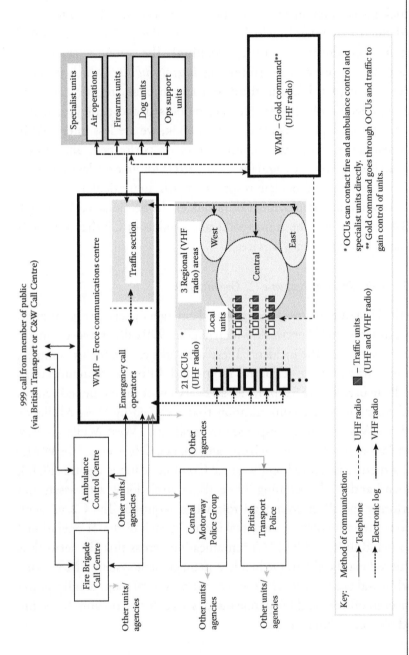

Figure 5.3 Schematic of emergency response in a regional control in a UK police force. C&W, Coventry and Warwickshire; UHF, ultra high frequency; and VHF, very high frequency.

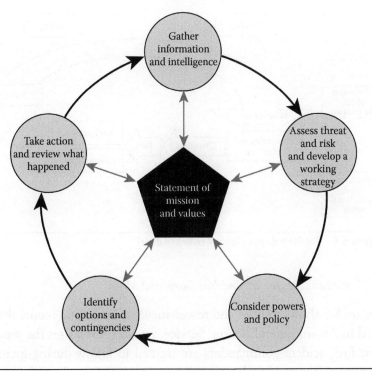

Figure 5.4 NDM.

The national decision model (NDM) is a generic process for the assessment and response to any police incident; it is widely used throughout police services for both planned and unplanned events. Whilst the NDM is viewed as the key framework for operational planning, it may also be used to assist with decision making at strategic and tactical levels.* Figure 5.4 presents the National Decision Model as an iterative process for gathering information, making assessments, planning and then acting upon the environment. Information is used to inform threat assessment which then leads to consideration of appropriate actions to take (in terms of powers and policy and available options). In a similar manner to the data-frame model (Section 2.3) the process is assumed to be cyclical and iterative, so that decisions are continually monitored and, if necessary, updated.

---

* National Police Improvement Agency (NPIA). 2010. *Manual of Guidance on Keeping the Peace.* London: National Police Improvement Agency.

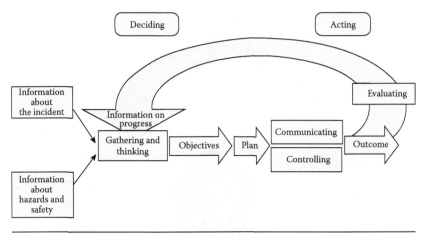

Figure 5.5 The fire and rescue incident command model.

5.3.2 Fire and Rescue: The Incident Command Model

Figure 5.5 shows the fire and rescue incident command model that is used in Avon Fire and Rescue Service, which summarises the process that fire incident commanders are trained to follow during incident responses. The model identifies two main activities, where the stages to the left of the model fall into the *deciding* activity, whilst the stages to the right form the *acting* phase of the cyclical process.

Both the police (NDM) and fire and rescue command models are essentially describing the same cyclical process; although the fire and rescue model appears more complex than the NDM, this is mainly because it makes explicit a number of activities that are implicit in the NDM. These models also call to mind the Boyd Cycle, or the *observe, orient, decide, act* (*OODA*) loop, which typifies an approach that has been prevalent in military C2 thinking since the 1970s (Figure 5.6) (Boyd 1996).

5.3.3 OODA Loop

Some of the main themes associated with the OODA loop are as follows:

- Performing OODA more effectively than the adversary means that you can gain the advantage during conflict.
- Orientation is a highly complex process, made up of a number of factors, including cultural factors and previous experience.

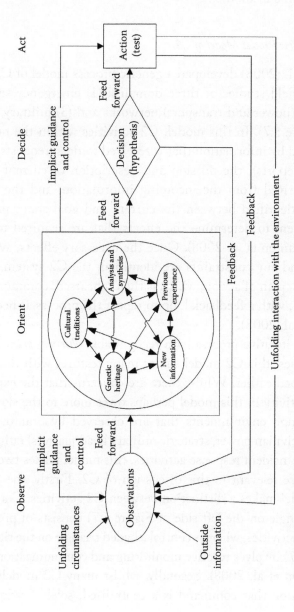

Figure 5.6 The Boyd Cycle/OODA loop in full detail.

- Orientation influences observation, as well as deciding and acting.
- All processes are taking place at once and with multiple influences on one another.

5.3.4 A Generic Process Model of C2

Stanton et al. (2008) developed a generic process model of C2 from a series of field studies of three domains: (1) emergency services, (2) civilian (power and transport) networks and (3) military operations (Figure 5.7). In this model, C2 activities within the network are triggered by information that is received (orders, requests, intelligence or reports); the mission and description of current events are then derived from the incoming information, and the subsequent gap identified between the current and goal states prompts the C2 system to determine the effects that are required to close this gap (Stanton et al. 2008). Once the necessary effects, available resources and any constraints are identified, the C2 system is able to generate a plan of action, which is then rehearsed, communicated and enacted, with a feedback loop to repeat the process as necessary (Stanton et al. 2008).

Being an iterative process with recognisable *observe, orient, decide* and *act* phases, this C2 model shares many features with those that were described earlier. Whilst there are concerns that the expanded planning activity in this model perhaps owes more to the slow- and medium-tempo environments that are observed by Stanton et al. (2008) (i.e. civilian power, strategic military headquarters) rather than high-tempo incident response activities, this model makes two assertions that are relevant to this discussion of C2. Firstly, the generic process model makes a distinction between C2 activities. Command (shaded triangle on the left side of Figure 5.7) consists of proactive, goal-based activities, whereas control (shaded triangle on the right side of Figure 5.7) involves reactive monitoring and communication activities (Stanton et al. 2008). Secondly, whilst many C2 models make the assumption that command is a centralised, solely headquarters

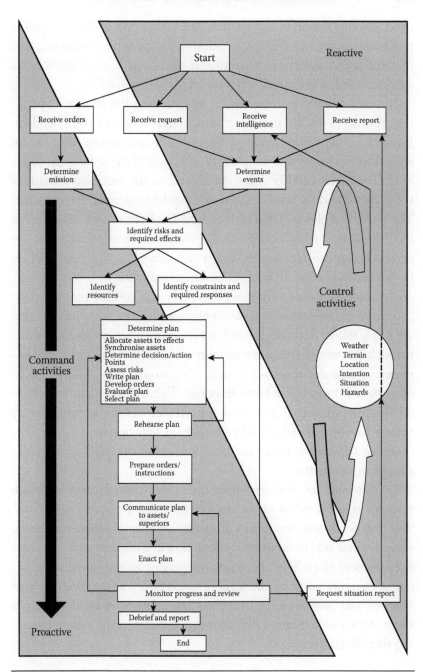

Figure 5.7 Generic process model of C2.

function, this model recognises that the ownership of the various C2 process elements can change, depending on the type of network, and that the decentralisation of command can occur '...*right down to the level of personnel in the field*' (Stanton et al. 2008, p. 233).

In many situations, command is enshrined in the standard operating procedures (SOPs) that are enacted in response to a given incident. From this perspective, sensemaking could be seen as the definition of an incident in terms that support the selection of the most appropriate SOP. The choice of SOP will then depend on the *sense* that is made of the incident and the resources that are available to make the response. The two examples in Chapter 1 might imply that such selection of SOP is performed solely by the dispatcher. This is not the case for several reasons. First, the dispatcher will consult with a senior officer in the control room. Indeed, dispatchers are often in a civilian role, and a ranking police officer will work in the control room to take command of incident management and to authorise the decisions that are being taken. Second, dispatch is becoming increasingly automated through the use of computer-support systems that can either recommend a particular response or that can encourage the interpretation of an incident towards a given response. Third, the responding officer will interpret demands, risks and opportunities at the scene and share these with the dispatcher and colleagues as appropriate.

This could lead to either the modification of the SOP or the selection of an alternative.

In this manner, incident response constitutes control in which resource is allocated to an incident in line with an appropriate SOP. This implies a reactive approach to response in which sensemaking plays a minor role in defining the situation in sufficient terms to allow the selection of an SOP. However, the examples presented in subsequent chapters point to the level of uncertainty and ambiguity that incident response typically involves, and show how sensemaking is more often a proactive act of managing resource in the face of changing situational demands.

5.4 The Future of C2

Attempts to rethink C2 have focussed on designing organisations to be more agile, i.e. to be more resilient, efficient and responsive to sudden changes (Alberts and Hayes 2003). Alberts and Hayes (2003) describe the edge organisation where individuals on the front line are empowered to act as circumstances change, rather than having to wait for orders to be issued. This would be done by giving individuals access to all relevant information and expertise and through the removal of procedural constraints. This notion of distributed, *ad hoc* groups acting autonomously and asynchronously is in stark contrast to the more regimented, synchronised and deconflicted operations that are associated with traditional hierarchical C2.

One approach to providing the conditions for agility is to fully connect all agents within the network, thus radically altering the organisational structure away from the traditional centralised hierarchy. It is contested that such change would resolve the long-standing trade-off between the *reach* and *richness* of information, i.e. groups of individuals no longer require physical proximity in order to be able to share detailed information quickly (Evans and Wurster 2000). Technology support for such change could involve information management systems that enable widespread sharing of information and the creation of task-focussed communities of interest that develop in order to collaborate on specific issues that arise. Against this, we have previously cautioned against the assumption that *'the structure and behaviour of a network are equivalent'*, i.e. that changes to one will automatically lead to changes in the other (Baber et al. 2008). From this perspective, the structure of the organisation could influence its sensemaking behaviour.

6
SENSEMAKING IN COMMAND AND CONTROL

Two generic forms of command and control (C2) network structure (community of practice versus networks of exploration) are compared, in terms of the ways in which the members of these networks collaborate, the ways in which planning is performed and the ways in which problems are handled and response is coordinated. The argument is that different phases of a response exhibit characteristics of the two network structures. By considering how tight and loose coupling of members of the networks can lead to different challenges for managing activity, we consider how the network structure has a bearing on the ways in which the sensemaking is achieved and the ways in which sense can be shared.

6.1 Introduction

Sensemaking relates to the concept of *having the bubble*, i.e. maintaining a big-picture view of operations, which is seen as being of crucial importance in safety-critical environments (Roberts and Rousseau 1989). The ability of distributed networks to engage in complex and fast-paced activity means that they will move beyond the ability of any single individual to maintain a detailed picture of what is going on. However, complexity (including complex and distributed decision making), tight coupling and fast pace of action do not necessarily lead to organisational failure, as is demonstrated by high-reliability organisations, such as aircraft carriers (Roberts and Rousseau 1989). *The bubble* (more likely a series of nested bubbles) and, consequently, safe operations are maintained in this environment through clear roles, responsibilities and lines of command; high levels of training and established processes for every aspect of operation; it is questionable, whether within high-tempo networked operations that feature *ad hoc* reorganisation and decision making *on the hoof* at the edge as core principles, that systemic failures can be avoided.

An example of the potential type of failure was provided during Operation Telic in 2003, by the unplanned decision of two A-10 pilots to engage targets of opportunity, which were subsequently revealed to be British armoured personnel carriers (Townsend 2007). During the cockpit video recording, the two pilots can be seen to retrospectively construct a narrative frame that supports their decision to engage the vehicles; this decision goes unchallenged by the wider organisational network until it is too late. This caveat does not mean that co-located teams operating in complex environments are guaranteed to enable personnel to have the bubble; Rochlin (1991) discusses how the increasing technological sophistication of US warships led to a reduction in the awareness of personnel and contributed to the shooting down of Iran Air Flight 655 by the USS Vincennes in 1988. This illustrates the requirement to consider organisational change from a sociotechnical perspective in order to prevent unanticipated consequences.

In the discussions of networked operations, the term *network* is often interpreted as *networking technology*, with the inference that social and organisational aspects of the system will then fall into line (c.f. MoD 2005). Previous research into C2 would suggest that this is not the case and that a more sociotechnical systems view – i.e. the optimisation of both social and technical systems (Parker and Wall 1998) – should be taken during the design and implementation, in order to avoid unintended consequences and thereby worsened performance (McMaster and Baber 2006; McMaster et al. 2006). To this end, it is important to have an understanding of how the use of distributed networks may have an influence on C2 processes such as information processing and collaborative (team) activities, such as problem-solving and decision making. In fact, work redesign research informed by sociotechnical systems theory has already been carried out on autonomous work groups, distributed networks, computer-supported collaborative working and a number of other group processes that are featured within networked operations, so a review of the relevant literature can begin the process of identifying important issues in relation to this area.

6.2 Collaborative Networks

Fraher (2011) criticises the centralized C2 paradigm, where communication is seen as the *'exchange of "information" and "instructions" from one leader to recipients'* (p. 181), advocating decentralized, exploratory structures instead, where the leadership role may be shared, and sensemaking is a collaborative process. Rather than seeing C2 structures as a choice, we believe that both centralised and decentralised structures can be useful, but the structure used could depend on the nature of the incident. Assuming that response systems can be loosely or tightly coupled (see Table 6.1), one can appreciate how the routine incident response described in Chapter 1 broadly fits the characteristics of tightly coupled systems, whereas the major incident response bears a closer resemblance to the description of loosely coupled systems. This suggests that the incident type might result in different forms of coupling in the C2 system.

Table 6.1 Tightly and Loosely Coupled Work Systems

KEY DIMENSIONS	TIGHTLY COUPLED SYSTEMS	LOOSELY COUPLED SYSTEMS
Access to resources	Agents and representational artefacts are restricted to a predetermined set.	Agents and representational artefacts are unrestricted to a predetermined set and may change over time.
Problem structure	Well-structured, identifiable and expected problems that are recurrent.	A tendency toward ill-structured problems that have a high degree of uniqueness
Organisational structure and problem dynamics	Organisation has pre-specified modes of operation. Division of labour is well understood, and *standard operating procedures* underpin much of normal work.	Organisation's operation is only partially predetermined; established work processes are augmented by *ad hoc* approaches. Divisions of labour are informally defined and enforced.
Cycle duration	Relatively short cycle for problem-solving, coupled tightly to the task.	Problem-solving cycle tends to be variable.

Source: Perry, M., Socially distributed cognition in loosely coupled systems, in S. J. Cowley and F. Vallée-Tourangeau (Eds.), *Cognition Beyond the Brain: Computation, Interactivity and Human Artifice*, London, Springer, p. 162, 2013.

Table 6.2 Potential Impacts of Different Command Arrangements on Macrocognitive Processes

	CO-LOCATED	DISTRIBUTED
Traditional hierarchical C2	*Sensemaking*: in the bubble; command oversight. *Planning/adaptation*: adaptable command structures. *Co-ordination*: wider information-sharing options. *Sensemaking*: groupthink. *Planning/adaptation*: influence of status; vulnerability limits adaptation.	*Sensemaking*: in the bubble; command oversight. *Planning/adaptation*: adaptable command structures; resilience to disruption. *Problem detection*: detection of subtle/diffuse cues. *Problem detection*: fallacy of centrality; difficulty communicating cues. *Co-ordination*: lack of shared experience.
Edge networks	Communities of practice[a] *Problem detection*: detection of subtle/diffuse cues. *Co-ordination*: wider information-sharing options. *Sensemaking*: groupthink *Planning/adaptation*: influence of status; vulnerability limits adaptation. *Problem detection*: noisy environment masks cues; fallacy of centrality.	Communities of practice[a] *Planning/adaptation*: resilience to disruption. *Problem detection*: detection of subtle/diffuse cues. *Sensemaking*: systemic failures; lack of command oversight. *Planning/adaptation*: problems adapting. *Problem detection*: noisy environment masks cues; fallacy of centrality; difficulty communicating cues. *Co-ordination*: lack of shared experience.

[a] Burnett et al. (2004) contrast the concept of the *community of practice* (c.f. Wenger et al. 2002) with their concept of *exploration network*.

Table 6.2 contrasts hierarchical with edge networks in C2 in terms of whether decision making and sensemaking is distributed or co-located. In this table, a comparison is made in terms of sensemaking, planning and adaptation, problem detection and co-ordination. This distinction is explored further in Chapter 7.

When confronted with an uncertain, ambiguous or novel incident, the initial response might be most effectively undertaken with an exploration network to gather information from as many sources as possible and to have a loosely coupled command structure to allow a highly flexible response (Table 6.3). And as the incident becomes better defined and understood, then it becomes more efficient to enact standard operating procedures through a community of practice and to move to a tightly coupled response system (Baber et al.

Table 6.3 The Characteristics of Communities of Practice and Exploration Networks

COMMUNITY OF PRACTICE	EXPLORATION NETWORK
Specialized terminology.	Everyday language.
High levels of abstraction.	Low levels of abstraction.
Shared practice and domain of interest.	Shared experiences, values and beliefs.
Well-defined practice within the domain – the set of frameworks, tools, information, language and documents that the community shares.	The development of a practice is a possible, long-term outcome of exploration, not a given.
Well-defined areas of common interest (the domain of the community).	Often poorly defined areas of common interest.
Long-lived, relatively static membership.	Short-lived, dynamic associations.
Community members defined by professional or organizational groupings.	Networks form and re-form depending on task and need.
Goal is incremental improvement in applying knowledge in a well-defined area.	Goal is to develop new interpretations, conjectures, ideas and ways of looking at the world that may be exploited for a purpose.

Source: Burnett, M. et al., Sense making – Underpinning concepts and relation to military decision-making, *9th International Command and Control Research and Technology Symposium*, Copenhagen, Denmark, September 14–16, Washington, DC, CCRP, p. 13, 2004.

2007). This suggests that the way in which sensemaking is performed could change between these different structures. In the exploration network, sensemaking becomes a matter of working through uncertainty, experiencing gaps in understanding and seeking to fill these gaps in order to make sense (either individually or collaboratively). In the community of practice, sensemaking becomes more a matter of ensuring continued coherence of understanding, perhaps dealing with novel information that might require an adjustment or refinement of the current *sense* that is being used to guide response.

6.3 Planning and Adaptation (Replanning)

Planning in C2 takes place at a number of levels, from long-term strategic decisions to the more time-pressured operational and tactical levels. Dubrovsky et al. (1991) found that during face-to-face (FtF) planning and decision-making discussions, proceedings tend to be dominated by high-status individuals. This could result in a failure to consider all relevant information or to explore possible options thoroughly. The use of electronic communications appears to reduce this effect (Dubrovsky et al. 1991), suggesting that distributed teams may be more egalitarian in discussions and decision making.

Adaptation, or agility (Alberts and Hayes 2003), relates to resilience, efficiency and responsiveness. In relation to macrocognition, adaptation relates to the ability to engage in replanning, i.e. the modification or replacement of a plan (Klein et al. 2003). Replanning itself is reliant on some form of feedback from the environment, which clearly involves the associated functions of sensemaking and problem detection; however replanning also relates to a team's ability to adapt to varying task demands and openness to new perspectives (c.f. *fallacy of centrality* on Section 6.4 of this report). Serfaty et al. (1993) found that teams with records of superior performance appear to be very adaptive to varying task demands – moving between different organizational structures, communication patterns and roles as the nature of the situation changes. They hypothesised that the reason for this is that co-ordination strategies evolve from explicit co-ordination under low-workload conditions to implicit co-ordination as the workload increases. Whilst distributed networks have been labelled as *agile* by some proponents, it may be the case that such edge organisations will struggle to switch between different patterns of work to suit the circumstances.

If, as Serfaty et al. (1993) suggest, it is the case that adaptability evolves from explicit co-ordination under low-workload conditions, then within fast-paced edge organisations, where groups form on an *ad hoc* basis in order to work on specific tasks, there would not be an opportunity to work together under low-load conditions and enable alternative modes of working to develop. The distributed nature of co-operation within such organisations is likely to pose an additional hurdle for implicit co-ordination strategies to develop. The possible implication of this is that whilst distributed teams may be able to function at a fast pace under *normal* conditions, once the nature of events reaches a point where different modes of operation are required, such teams may not be able to adapt sufficiently to continue to function effectively. Potter et al. (2007) describes a method for supporting adaptation within dynamic distributed work environments, but this process requires a level of stability within the organisation in order to build a model and identify the factors affecting system performance; stability is the antithesis of the *edge organisation*, indicating that there may be a practical limit to the amount of *ad hoc* flexibility, which will be possible within distributed command networks.

The description of adaptive teams given by Serfaty et al. (1993) seems to match with the emergency services' ability to move between their standard response C2 – in which much of the *command* element is latent – and a rapidly expanded organisational structure in response to large-scale emergencies (Baber et al. 2008). Whilst this structure has been developed and trained, rather than evolving naturally, and undoubtedly has a limit as to the pace of events with which it is able to cope (c.f. McMaster et al. 2007), it does serve to illustrate that distributed edge networks are not automatically the superior choice in all circumstances.

Whilst the notions of vulnerability and resilience that are sometimes associated with co-located and distributed networks are not obviously relevant to cognitive functions, there are some indirect impacts on an organisation's ability to react to changing circumstances, particularly where an agency is based and operates within a specific geographic area – for example, incident response organisations. The Emergency Operations Centre (EOC) for New York City (located in World Trade Center building 7) boasted a modern facility for the co-ordination of responses to any emergency that the city might face, having a workspace and computer equipment for up to 68 agencies, sophisticated communications networks and access to live cameras and databases relating to all aspects of the city (Kendra and Wachtendorf 2003). However, on September 11, 2001, the attack on the World Trade Center complex forced the evacuation of the EOC and the relocation of personnel to several temporary locations, though the subsequent expansion of the hazard area rendered these sites inoperable. A suitable location was eventually found, and information technology and office equipment was installed within 36 hours (Kendra and Wachtendorf 2003), though the intervening disruption would undoubtedly have hampered the disaster response effort. The concentration of resources within a centralised co-located command system means that the command structure itself can become a liability; in addition to the disruption caused by an unplanned relocation, the sudden discovery that the *hub* of a centralised command system is itself within the hazard zone of a serious incident is unlikely to benefit decision making, as there is no longer a psychological distance from the incident in question, and commanders are forced to consider tactical issues surrounding the preservation of the command hierarchy, rather than remaining detached and focussed on wider strategic considerations.

In contrast, within geographically distributed networked organisations, there is less likely to be a sudden compromise of the entire decision-making group, and where problems develop in specific areas, the self-organising and *ad hoc* nature of the problem-focussed team means that more individuals can be brought in to continue to work on the problem (Weick et al. 2005).

6.4 Problem Detection

Problem detection is the term that is given to the cognitive process by which individuals become aware that the course of events may have taken an unexpected and undesirable direction that may require action (Klein et al. 1999). Failure to detect problems sufficiently early can result in degraded performance, and the system may eventually deteriorate to the point where recovery is impossible (Klein et al. 2005); within the military domain, such loss of performance is likely to have disastrous consequences. Within the macrocognitive framework, problem detection is thought of as a reconceptualization, or reframing of the situation, rather than the traditional view, which was that discrepancies progressively accumulate until a threshold is reached (Klein et al. 2005). As such, problem detection is a form of sensemaking that relies on the recognition of subjective (not only to the situation but also to the individuals who are involved) cues that throw doubt on the way that the situation is being framed (Klein et al. 2005).

Klein et al. (2005) emphasise the importance of expertise in identifying subtle cues, and it may be the case that teams within networked distributed organisations have an advantage in this respect, through the ability to rapidly recruit relevant personnel from across the organisation to attend to a specific activity. Additionally, the ability of distributed teams to rapidly collect and collate information may enable them to identify signs of a problem through a range of small, diffused cues, particularly in situations of high uncertainty, where information is missing, unreliable or conflicting (Klein et al. 2006b).

Problem cues may be masked by other problems or through background *noise* within the environment (Klein 2006b); fully networked, distributed teams feature very high volumes of information traffic, and whilst steps may be taken to try to pick up the most important cues, there is the danger that they may be lost in the flow of information. Klein et al. (2005) discuss the tendency of people to explain away information that contradicts their frame of the situation. Westrum (1982) refers to the fallacy of centrality, whereby experts – feeling well placed within their information networks – overestimate how much they would know about an event if it was actually taking place. Westrum argues that this fallacy is doubly damaging because it discourages curiosity on the part of the individual and predisposes him or her to take an antagonistic stance towards the event (Westrum 1982). Weick (1995) then extends this observation to argue that networked organisations may experience similar problems, whereby new information (for example, a problem cue) may be discounted because individuals conclude that it is not credible; otherwise, they would have known of it sooner. Problems surrounding the fallacy of centrality are further compounded by the fact that expectations about what one would know are based on tacit assumptions, which are rarely articulated or questioned (Westrum 1982). In fact, it could be argued that these assumptions at least partially underlie sensemaking and therefore problem detection.

The fallacy of centrality could be said to have arisen within networked military operations even before they became a reality – notions of *total information awareness* and *information superiority* have been discussed, particularly within US military circles, for a number of years and would appear to chime with Westrum's (1982) observations about overestimation of the effectiveness of information-sharing processes (Popp and Poindexter 2006). Teams within distributed-edge organisations, which purport to give them access to all the information and expertise that they require to make effective decisions, must be especially mindful of the tendency to believe that this is the case, when in reality, in complex, rapidly changing and highly uncertain environments, it will never be possible to be in possession of all the facts. The problem of friendly fire incidents was not only discussed in Section 6.1 in relation to sensemaking, but it also clearly relates to problem detection and the fallacy of centrality and so should be of particular concern to networked organisations.

6.5 Co-Ordination

Given that this section is concerned with the group performance of macrocognitive functions, it is unsurprising that co-ordination – a fundamental aspect of group work – is associated with the performance of the other functions that are under discussion; in addition, co-ordination of activity clearly requires elements of joint sensemaking, planning and decision making. The link between co-ordination and adaptability identified by Serfaty et al. (1993) was discussed in Section 6.3; another example is provided by Jentsch et al. (1995), who found that within aircrews involved in flight simulations, crew co-ordination behaviours significantly predicted the time that is required to identify a problem with the flight. A range of active and passive team co-ordination behaviours have been identified by researchers, as well as explicit and implicit modes of co-ordination (Stone and Posey 2008). Stone and Posey (2008) examined what effect computer-mediated communication (CMC) had on team co-ordination over FtF interaction; they found that different behaviours (namely, help seeking or inquiry behaviours) were predictive of performance for the CMC group rather than the FtF groups. Stone and Posey (2008) interpreted this as indicating that because the CMC group members have reduced the monitoring of each other, they need to seek help or inquire about aspects of the situation, whereas FtF groups can more easily rely on monitoring and implicit communication (i.e. acting or sharing information without being asked to do so).

Co-ordination issues for distributed teams have implications for other macrocognitive functions, including problem detection; Klein (2006b) identifies a number of problem-detection issues that teams may face, such as a failure by the team to realise that a common understanding has been lost, or difficulty in communicating the urgency of a perceptual cue – both of which are likely to be exacerbated by the reduced monitoring and availability of implicit co-ordination that is experienced by distributed teams.

Co-ordinated action requires some form of equivalent understanding (i.e. sensemaking) amongst participants; however, researchers have argued that in work situations, whilst people share actions, activities, conversations and joint tasks, they do not share meaning (Donnellon et al. 1986; Weick 1995 – all cited in Brown et al. 2008). This leads to the problem that was stated in Section 6.5 that group members may have significant sensemaking discrepancies (Brown et al. 2008). Weick (1995) proposes that despite these discrepancies, co-ordination is still possible because individuals within the group share a common experience that ties them together, despite the different meanings that they may infer from it. Van Fenema (2005) argue that within the high-reliability organization domain (emergency services, military operations, air traffic control, etc.), the tendency to operate in a geographically dispersed manner contributes to the complexity of the system and thereby to the level of risk. They present an analysis of a mid-air collision, determining that the failure of distributed co-ordination was due to a number of factors, including gaps and differences in knowledge that are held by different agents within the system (i.e. sensemaking discrepancies) and technical interoperability issues. They conclude that time pressure places limitations on distributed co-ordination; another interpretation may be that by distributing co-ordination across geographically dispersed systems, there is no longer a shared experience for those who are involved to use as a common point of reference, increasing the risk that co-ordination activity may break down altogether.

6.6 Conclusions

Routine response is a matter of *control* (in contrast to command), in terms of coordinating resources in response to an emergency. Here, sensemaking could be seen as the activity that is a precursor to control decisions. In major incidents, there is a need to combine command intent (in terms of agreeing what situation is being tackled) with control. In this case, sensemaking could be seen as an iterative process of interpreting the situation and defining an appropriate response. As the situation changes, the need for sensemaking is to allow the response to adapt to such changes. Relating this need for adaptation means that the command structure needs to ensure that the right people *have the bubble* at the right time. This, in turn, affects problem detection and co-ordination, with problems arising from the fallacy of centrality in which the central command might assume that it is well placed to interpret and understand all aspects of the situation. The next chapters consider the examples of managing routine and major emergencies.

7
Managing Routine Incidents

Chapter 7 presents examples of incident management, predominantly based on research that is conducted with Warwickshire and West Midlands Police (WMP) forces in the United Kingdom. The process is broken down into four phases: (1) making sense of the call; (2) supporting responding units; (3) officer attending and (4) closing the incident. Each phase has its own opportunities and challenges for making (and sharing) *sense* of the incident.

7.1 Introduction

As we noted in Chapter 1, the response to routine emergencies begins with an often-fragmentary and inaccurate account from a distressed member of the public. The task for call handlers, police officers and other staff is to match this account to a recognisable policing framework, identify the appropriate response and the most suitable resources and then verify or reframe the incident in order to resolve the situation.

In this book, we employ the term *routine* to incidents that are regularly encountered, and staff are familiar with dealing with them, following standard procedures. As such, whilst there is command oversight, the organisation is largely capable of responding to these incidents with minimal direct command involvement.

Typical examples of *routine incidents* include the following:

- Burglaries in progress
- Criminal damage (including arson)
- Domestic violence

- Medical emergencies (including suicidal and acute mental health problems)
- Retail thefts
- Road traffic incidents
- Serious assaults
- Street robberies
- Urgent welfare concerns (e.g. elderly and disabled persons collapsed in their homes)
- Vehicle crime (e.g. theft from, theft of and driving offences)

These different types of incidents present a range of challenges, the uppermost of which is managing the risks to the public and responding officers. Consequently, the type of incident and any contextual factors will dictate the approaches that are used to respond to them (Flin et al. 2007). Ensuring that an appropriate response is put in place is therefore the primary concern of the emergency service command and control (C2) system, and given the level of uncertainty that surrounds emergencies, making sense of the unfolding incident is crucial to identify the correct response. This echoes Alberts and Hayes' (2003) C2 capabilities, which were described in Chapter 5. Thus, the purpose of incident response C2 is to detect and make sense of unfolding incidents, in order to put in place appropriate responses to minimise or avoid loss or harm, and then to monitor and direct the resolution of the situation, co-ordinating with partner agencies where necessary (Figure 7.1).

Despite the designation of 999 as for emergency use only, many calls to this number relate to non-urgent matters, for example, where the crime is historical or where the caller's problem is not within the remit of the emergency services. Such incidents are not dealt with in this book.

Returning to Stanton et al.'s (2008) generic process model of C2, routine incident response (Chapter 5) could be viewed as largely representing *control* activity. If *command* is interpreted as meaning *commander's/command intent*, then this is largely latent within the system during routine emergencies – goals are those as stated for each emergency service (e.g. *protect life*), as well as particular local priorities (as articulated in shift briefings); resources are those response officers on duty; and constraints include the legislative framework, the standards of performance and other ongoing incidents. Plans are

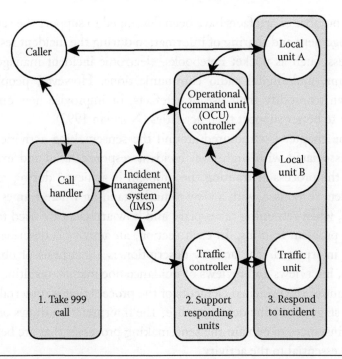

Figure 7.1 Routine incident response tasks, personnel and lines of communication.

largely standard responses to defined incident types. Whilst monitoring of district-wide events takes place (shift sergeant, control room supervisor, Force Communications Centre duty inspector, operational command unit [OCU] duty inspector), the vast majority of routine emergencies are resolved without any command-level input. Operational command-related activity (i.e. determining the mission; identifying resources; determining, communicating and enacting the plan) is instead carried out by the C2 network agents who are directly involved in co-ordinating the response (i.e. call handlers, controllers and response officers). This is similar to the military notion of mission command, whereby subordinate command levels are briefed on the goal of the mission but are then given freedom to determine how best to achieve their assignment. Consequently, *control*-level activity is a valid focus for research that is intended to investigate sensemaking during routine incident responses. Similarly, as several artefacts have been designed or adapted to support individual and collaborative incident response activities, they are a logical focus of attention for research into sensemaking within the C2 system.

A number of artefacts have been developed to support the capture, arrangement and sharing of information during the incident response process, such as pocket notebooks, electronic incident management systems, and digital radio communications. However, people are known to modify their use of artefacts, or improvise new ones, in order to better support their activities (Norman 1993).

The chapter is structured around the sensemaking activities that are associated with high-level incident response tasks and explores how the process of framing the problem is achieved during routine incident responses, with a view to understanding where frames come from, when reframing takes place and how artefacts are used to help with problem framing. In each section, an activity is described and then interpreted. The activity description is a synthesis of observations, interviews and reviews of guidance documents, resulting in an account of the principal elements of the process that is undertaken in each stage of the incident response. The interpretation draws on specific instances to explain the sensemaking processes that are believed to be essential to the activity.

Chapter 1 described how *routine emergencies* are regularly encountered, and staff are familiar with dealing with them, following standard procedures and generally with minimal direct command involvement. Consequently, in conducting this research, the expectation was that during routine incident response sensemaking activities, the police C2 network would be revealed to function as a community of practice, that is, as an established organization of individuals operating within a well-defined domain and that '...*share a common set of patterns of interpretation, implicit assumptions, and beliefs...*' (Burnett et al. 2004, p. 12).

7.2 Making Sense of the Call

First, 999 calls are answered by an operator who will ask the caller to specify an emergency service; once a service has been selected, the call is then passed on to a call handler within the geographic area where the call originated. The call handler's role is to gather details from the caller and establish the nature and severity of the incident. Incident details are entered into an electronic log, which is then passed to a dispatcher (generally known as controllers, or control), who are either in the same control centre, or distributed across local control rooms, depending on the structure of the force.

As an example of how the activity is supported by technology, the call handler uses the call-handling software (labelled '**1**' on Figure 7.2) to answer the call. Answering a 999 call causes the electronic Incident Management System (IMS) to automatically open a new log (**2**) and populate some of the log fields with information, such as date and time and the calling number and address (extracted from the telephone number, although this depends on the telecommunications company handling the call and whether the call is being made from a landline). The call handler will greet the caller with a phrase such as '*police emergency?*', prompting the caller to state the reason for his or her call. As the caller is speaking, the call handler may check that a log has not already been created for the incident (**3**) before noting down key details of the incident on their notepad (**4**).

Once the call handler has clarified the nature and urgency of the incident, he or she will restructure key details from the caller into a clear and concise summary of the incident, which is then entered into the IMS. The IMS requires that the call handler grade the priority of the call (e.g. *immediate, early, scheduled response*) and select from a defined set of incident types that are used to classify the nature of the incident.

Call handlers are trained to use a variety of question styles in order to direct the conversation and to establish the important facts quickly, including open, closed, alternative and leading questions (Warwickshire Police 2005). Open questions are often used to encourage callers to elaborate and are known as the 5Ws – *who, what, when, where, why* and *how*, e.g.

Figure 7.2 A call handler's workstation.

- Call handler: *'Have you been injured?'*
- Call handler: *'Where did they go?'*
- Call handler: *'Was it a male?'*
- Call handler: *'…and he got into a car?'*
- Call handler: *'Do you know if he was white, black or Asian?'*
- Call handler: *'What sort of age?'*
- Call handler: *'Lime green top…anything else?'*
- Call handler: *'Did you see the driver?'*
- Call handler: *'What sort of car?'*
- Call handler: *'What was in the bag?'*
- Call handler: *'What does the bag look like?'*
- Call handler: *'How large is it…what sort of material…?'*
- Call handler: *'Wait where you are. The officers are on their way'.*

The language used in the account may change, as call details are converted into abbreviations and police jargon. For instance, the description of an offender may change from *'white lad'* to *'IC1 male'*, which is the relevant UK Police National Computer (PNC) Ethnicity Classification. Abbreviations and acronyms are also employed, for example, *'My car has been stolen'* is formalised within the police as *'theft of motor vehicle'*, which is written as *'TOMV'*.

Figure 7.3 summarises the process of taking a 999 call in response to a robbery. The notation used in this figure was presented in Chapter 1 and is an annotated process model. The boxes on the right-hand side of the figure show the information that is recorded in the notepad and incident log at various points, showing how the incident log gradually develops during the course of the call. The figure also illustrates how the log structures the incident details and mediates indirect communications between the call handler and the controller. (The call handler can see that the controller has dispatched a unit to the incident and is able to tell the caller that the police will be with them soon.)

7.3 Supporting Responding Units

As they make their way to the incident location, officers begin to make sense of and plan their response to the situation. As the national decision model (Chapter 5) suggests, this includes considerations of risk (threat assessment), powers and policy and tactics. Although

MANAGING ROUTINE INCIDENTS 107

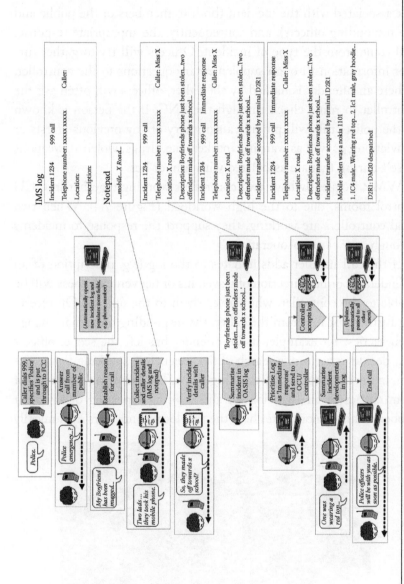

Figure 7.3 Annotated process flow for *taking a 999 call*.

the officers will have received some initial details from the controller, these are often only the bare minimum, such as an approximate location and a statement of the nature of the incident, for example, *'male being assaulted by two males'*. The first indication of the level of risk associated with the incident (both to members of the public and the responding officers), and consequently, the appropriate response, will come from the type of incident. Officers will try to gather further information through supplementary questions to the controller. Where an offender is named by the caller, officers will often ask the controller to run a check through the PNC; if the person is known to the police, this will provide a summary of any previous arrests or convictions, as well as warning markers (i.e. drugs, violence, weapons or self-harm), that are associated with him or her.

WMP frequently have two controllers working together in OCU control rooms, due to the high workload (Figure 7.4). When two local controllers are working, they support the response to incidents through very close co-operation.

If the call handler adds updates to the log (e.g. description of an offender, his or her direction of travel, his or her vehicle), these will be visible to the controller, who passes them to the officers. On receiving further updates from the caller, the responding units may change their tactics; for example, if the offender has left the scene, officers may decide to perform a search of the area before speaking to the victim, in the hope of quickly catching them. The following is an excerpt from entries in an IMS log for the ongoing reporting of a violent robbery. This demonstrates how the two local controllers dynamically

Figure 7.4 Controllers working together at adjacent workstations.

share both the radio talk group and the IMS log. Whilst one controller notionally takes a support role, both will broadcast over the radio, and each provides support to the other at various times.

05:19 Controller 1: '[Officer] *WITH 2 WITNESSES 1 OFFENDER IC1 MALE GREEN T*'
05:51 Controller 1: '*SHIRT THE OTHER IS IC1 MALE WHITE SHIRT AND TIE*'
05:57 Controller 1: '*BOTH MALES APPROX 20–25 YRS*'
06:09 Controller 1: '*THE OFFENDERS HAVE MADE OFF WITH HANDBAG AND METAL*'
06:17 Controller 1: '*TIN WITH LARGE AMOUNT OF CASH*'
06:24 Controller 1: '*LAST SEEN TOWARDS* [ROAD]'
...
14:20 Controller 1: '*THE IP HAS BEEN STRUCK AND FELL TO THE FLOOR*'
14:27 Controller 1: '*OFFICERS CHECKING TO ASCERTAIN IF AMBO REQUIRED*'
14:58 Controller 2: '*LADY HAS BEEN KNOCKED OVER AT DOOR WHEN OFFENDERS*'
15:00 Controller 2: '*GAINED ENTRY*'
15:15 Controller 1: '*CAN SOCO ATTEND ASAP PLSE*'
[Incident Switched for Scene of Crime Office (SOCO) tasking]
15:30 Controller 1: '*FROM OFFICERS THE FEMALE IP DOES NOT REQUIRE AMBO AS*'
15:32 Controller 2: '*PLS GET SOCO FOR THIS*'
[Incident Switched for SOCO tasking]
15:39 Controller 1: '*IP STATES HAS NO INJURIES*'
16:06 Controller 2: '*ASKING FOR AMBO ELDERLY FEEMAL BADLY SHAKEN APPROX*'
16:07 Controller 2: '*86 YRS*'
16:23 Controller 3: Incident Switch Accepted
17:36 Controller 1: '*THE OFFICERS NOW ASKING FOR AMBO AS THE IP 86YRS OLD*'
17:51 Controller 1: '*IS EXTREMELY DISTRESSED-UPSET*'
17:54 Controller 2: '*AMBO LOG* [Number]'
18:34 Controller 3: '*SOCO INFORMED*'

18:40 Controller 3: This incident added to SOCO list for section [Number]

The local radio talk group is used to support urgent transmissions (officer emergency assistance, emergency dispatch, immediate response co-ordination) as well as non-urgent communications (general announcements, officer-initiated database enquiries). In the following example, three overlapping discussion *threads* take place between officers and controllers (labelled 'A', 'B' and 'C') in rapid succession.

A   Mike 1: [Requests person check, gives details.]
    Controller 1: *'Not known'.*
    Mike 1: [Gives address]
    Controller 2: *'Not listed'.*
    Mike 1: [Gives a different name]
B   Officer: [Incident update: close, no crime]
    Controller 2: *'Received.'* [Updates and closes log]
    Controller 1: *'No exact match PNC – give me his postcode.'*
    Mike 1: [Gives postcode]
C   Controller 2: *'Any unit available for an immediate: reports of a male hitting a female at* [LOCATION]?'
    Mike 6: *'Mike 6: en route'.*
    Controller 1: *'Mike 1: safe to speak?'*
    Mike 1: *'Go ahead'.*
    Controller 1: *'He is wanted for* [OFFENCE]'.
    Mike 1: *'One in custody'.*

Figure 7.5 shows part of the *co-ordinate response* process for a break-in in progress incident, illustrating the role of the IMS and geographical information system (GIS) in supporting response co-ordination, with the IMS enabling direct communications between the geographically separate local and traffic controllers.*

In a similar manner to the process of *active listening* that is used by call handlers, responding officers will use what is known about an incident to cue frame-defined data collection, via a series of questions to the controller. For example, if there is a specific location for the incident (such as a residential building), officers may ask the controller

---

* This is the case for WMP; Warwickshire Police controllers are all based in the same control room.

MANAGING ROUTINE INCIDENTS

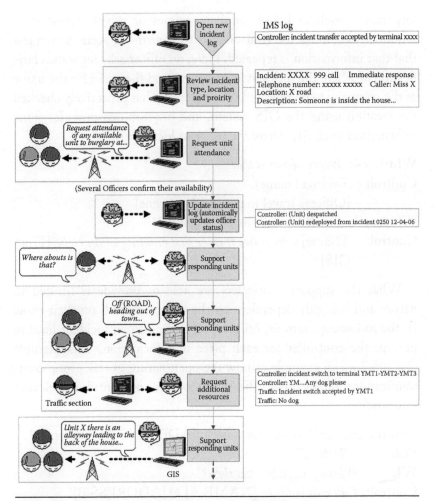

Figure 7.5 Annotated process flow for *support responding units*.

to check the IMS for previous emergency calls to that location, the details of any persons who are associated with that location and any previous convictions or warning markers (e.g. for violence or weapons) that are associated with those individuals. The artefacts available to the controller assist with this process, cueing controllers to provide updates to responding officers, which then cues further questions. For example, the IMS will indicate if the previous 999 calls have been made from a number or if any persons named in a log are associated with previous incidents.

The lack of mobile data access also means that officers will often rely on controllers to remind them of incident details that they have forgotten – such as house numbers, names or vehicle registration numbers – radioing the controller as they near the scene to request that that information is repeated. Here, an officer asks for some clarification of where the incident location was and then asks for the name of the company to be repeated. The controller has proactively checked the location using the GIS system, and unprompted provides some information to clarify where the incident location is.

Whiskey 4: *'Where abouts is that?'*
Control: *'Off* [road name]*'*.
[Officers travel to the incident scene]
Whiskey 4: *'TA. What is the company again?'*
Control: *'*[Name]*…looks like it is the first building on the right'* [From GIS]

What the support controllers are able to provide is limited in nature and is highly dependent on the other demands on their time. In the following example, officers responding to an incident have to prompt the controller for each piece of information. This example also provides an example of how the open nature of talk group communications allows other officers to monitor and contribute relevant information.

Control: *'Any unit for an immediate at* [ADDRESS A]*…Domestic?'*
W2: *'Whiskey 2'.*
W1: *'Whiskey 1…who lives there?'*
Control: *'Call was made by* [NAME A] *at* [ADDRESS B]*'.*
W1: *'I want to know who lives at this address'.*
Control: *'I haven't the faintest idea'.*
W1: *'Can you give us a clue?'*
Control: *'Wait one'.* [Starts to run checks on the address]
Sergeant: *'Whiskey 3-5: I believe it may be* [NAME B]*'.*
Control: *'Yes,* [NAME B] *and* [NAME C]*…have also had calls from that address by* [NAME D]*'.*

When other officers have knowledge that is relevant to an incident, they are often better placed to provide information than the controller. A woman dials 999 to report that a prowler in her garden has shone a torch in her window. Whiskey 2 and Whiskey 3–5 are dispatched as an immediate response. In this example, an officer who is not involved in the response but who is monitoring the talk group casts doubt on the current frame (incident type) and suggests an alternative based on experience, prompting the controller to search for corroborating information. This information influences how the attending officers will deal with the incident, as is indicated by their update of '...*we'll go and have a chat*'.

Officer A: '*Be aware:* [NAME] *lives there and calls the police every time a light goes on outside*'.
Control: [Checks IMS for previous calls]
Control: '*On STORM* [IMS] *there are 56 previous calls from* [NAME]...*well spotted*'.
Whiskey 2: '*Whiskey 2 – TA*' [Arrived]
[Whiskey 3–5 stands-down]
Control: '*Be aware: I believe this female put a complaint in about how her last call was dealt with*'. [from IMS checks]
Whiskey 2: '*Yes, she's stood in her doorway. We'll go and have a chat*'.
...
Whiskey 2: '*W2; the house is secure, it would appear they've seen someone shine a torch through the window. All in order. TL*'. [Leaving]

Control room observations revealed that collaboration between co-located controllers is not reliant on explicit communication but is rather based on reciprocal monitoring and a shared deep understanding of the task at hand. This raises a number of issues about the use of the IMS as a collaborative C2 tool:

- The IMS is sometimes used for explicit as well as implicit communications (e.g. making specific resource requests to the traffic controllers).
- Whilst the controllers are working together, they are not always aware of the information that the other has added (e.g. duplicate requests for SOCO and ambulance).
- Entries are highly compact, and the IMS is used to rapidly impart information between colleagues who share common ground, rather than to present a highly polished account of events to outsiders.
- The narrative of what has happened is gradually built up and refined over time (e.g. from an injured person having been

struck to having *fallen over*). This narrative continues for several pages and can be difficult to follow, even before it is fragmented by numerous automatic IMS event entries (e.g. when officers are dispatched, when they arrive, when a log is switched to a different user, when that user accepts the log).
- The IMSs used by the different emergency services are completely separate, so when controller 2 rings the ambulance service to request their attendance, they exchange incident numbers with the ambulance call hander so that the incident log can be identified if further calls are necessary.

7.4 Officer Attending

As they arrive at the scene, responding officers notify the controller (who updates the incident log); the officers may be confronted by an ongoing incident, or they may find that the immediate threat from the incident has stopped. Either way, in order to achieve the goals of restoring order, preserving life and property and the detection of offences and offenders (HM Inspectorate of Constabulary 1999), their response to the incident is concerned with two interrelated high-level tasks: (1) controlling and resolving the situation and (2) performing an initial investigation of the events surrounding it.

Where more than one officer is deployed to an incident, they may decide to separate and divide tasks between them (e.g. conducting searches, separating belligerent parties, speaking to witnesses), using their radios in point-to-point mode (i.e. direct one to one) to co-ordinate their activities without taking up airtime on the talk group.

Police officers are issued with a pocket notebook; this provides somewhere to record information when dealing with incidents, such as witness accounts, details of evidence and the officer's narrative description of events. Whilst the pocket notebook is principally for an officer's own use, the information recorded here will form the basis of subsequent investigatory paperwork. It may also be referred to by the officer when giving evidence in court (potentially years later) and therefore may be examined by lawyers or court officials. Consequently, there are strict rules governing when and how the notebook is used in order to support the reliability and accuracy of entries.

Officer enquiries are supported by the controller, who is able to verify details through police databases (e.g. vehicle registrations, names, addresses), check officer welfare, allocate additional units, contact other services, record updates in the IMS and circulate information to other officers (e.g. descriptions of suspects).

Responding to incidents is complicated by the fact that many of the incident details may well be inaccurate, including the caller's account of events, the names or descriptions of parties who are involved and, very often, the nature of the incident itself (i.e. the frame that is selected by the call handler during the initial call). Multiple units respond to reports of a break-in in progress at night; officers are on the scene within three minutes; however, upon their arrival, the property and surrounding houses appear to be secure and undisturbed, casting doubt on the nature of the incident. The controller switches the incident log back to the call handler (in a different control room) to double-check the address. The situation officers' encounter at the scene is at variance to the summary that they have been given, causing them to question the sensemaking framework for the incident. This, in turn, cues activity from the controllers and the call handler, who communicate with each other via the IMS.

12:46 Controller A: *'CAN YOU CONFIRM x RD OR x ST'*
13:00 Call handler: *'STANDBY'*
13:23 Call handler: *'I HAVE LISTENED TO TAPE AGAIN IT IS x STREET'*
13:28 Call handler: *'NOT ROAD – MY APOLOGIES'*
13:28 Controller A: [Receives no reply from caller's mobile phone]
14:10 Controller B: [Updates caller details to x Street]
14:16 Controller A: [Updates incident location to x Street]
15:20 Controller B: [Notes that the house numbers in x Street only go up to 12 – the caller had reported living at number 15]
15:50 Controller B: [Performs searches for the caller on the voters' database]
17:20 Call handler: *'I HAVE LISTENED TO ALL THE TAPES AND WHEN I CONFIRM'*

17:34   Call handler: *'THE NUMBER OF THE ADDRESS CALLER STATES x ROAD'*
17:44   Call handler: *'I REPEATED IT TO HIM AND HE SAID YES x ROAD'*
17:54   Call handler: *'AT THE BEGINNING OF THE TAPE HE STATES x'*
18:06   Call handler: *'STREET'*
21:50   Controller B: [Notes that officers have checked the front and rear of both 12 x St and 12 x Rd and spoken to resident at 12 x St – all in order]
24:50   Controller B: [All units are leaving the scene. The log is closed, having been redefined as a false call]

The pocket notebook and the digital radio appear to function as cognitive artefacts that support officers in making sense of the incidents that they encounter. Over the course of the participant-observation period of this research, it gradually became apparent that many officers had modified their use of the notebook, employing the back pages to make unstructured notes in a similar fashion to the call handler's notepad, which was described in Section 7.2. As the details of an incident become known during an officer's enquiries, they are often recorded in the back of the notebook; these details are used to cue further information gathering from witnesses, as well as the officer's actions, such as searching the area and questioning individuals who match the offender's description.

With support from the controller, officers often use their notebooks to their advantage while making sense at the scene of an incident. Figure 7.6 summarises the sensemaking process that is involved in verifying the identity of an individual. As with the call handler's notepad described earlier, the back of the pocket notebook also functions as a private cognitive artefact that supports frame seeking, by capturing key information that is divulged when questioning an individual. The person's details (i.e. name, date of birth, home address) are recorded in the back of the notebook and checked on police databases (via the controller over the radio). If these checks are negative, this cues the officer to question the frame and potentially provides an alternative one (e.g. this person is engaging in deception, possibly to conceal an offense), prompting further enquiries with the individual. The officer

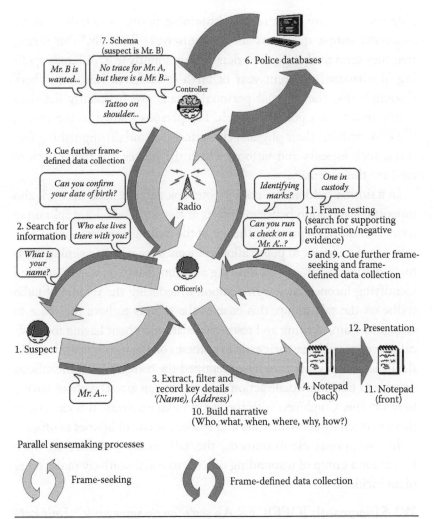

Figure 7.6 The sensemaking process involved in establishing the identity of a suspect.

may ask the individual to repeat his or her details and provide further information, such as other persons living at an address (who would be listed on the electoral register). The pocket notebook supports this reframing of the incident, providing a record of the earlier responses from the suspect and revealing any inconsistencies. By repeating this process of information gathering and database checks, the individual's deception becomes apparent, and officers will then take appropriate action.

At the same time, the officer and the controller are also able to take steps to establish the individual's true identity. Individuals trying to hide their identity must rapidly improvise fictitious details in order to answer simple questions about themselves. Typically,* this means that they tend to change their details only slightly (e.g. different spelling of surname, different year of birth or different house number). Consequently, their actual personal details are frequently listed in search results as a possible match. For persons known to the police, the PNC will list their physical characteristics and distinguishing features, such as scars and tattoos, which the officer can then check to confirm their identity.

In a similar fashion, the pocket notebook and the radio network also support collaboration between officers at the scene who have split up to conduct enquiries separately. The use of the radio as a private shared cognitive artefact (point to point) allows officers to engage in a similar iterative sensemaking process, comparing accounts (based on their notes) and identifying inconsistencies. In addition to reducing the volume of radio traffic on the talk group, this enables the officers to have the space to engage in frame seeking and testing/validation without having to physically meet or publicise their early hypotheses on the radio network and in the incident log. Once they have identified the correct frame, the officers are able to take immediate action (e.g. making an arrest) without having to reconvene. Combined with the use of radio earpieces, this can enable them to catch suspects unaware and reduce the risk of injuries to officers.

In exceptional circumstances, the talk group becomes an open forum for a group of responding officers to collaboratively make sense of an incident.

W3-5 [Sergeant]:	'[OFFICER A]: *you're on the wrong side…Unit looking at me, go down there'*.
Dog handler:	*'You were calling me?'*
Control:	*'Possible break-in in progress…'* [Gives details]
Dog handler:	*'Can you confirm I'm required?'*
Whiskey 3-5:	*'Yes – confirmed break-in'.*
Golf 3:	*'Golf 3: TA'.*
Control:	*'TA'.*
Whiskey 2:	*'Whiskey 2: What's the address again?'*

---

* This was observed by the researcher on several occasions and was corroborated in Subject Matter Expert (SME) interviews.

MANAGING ROUTINE INCIDENTS

Control:	'[ADDRESS]'
	[Confusion ensues over the location of the road and property]
Control:	*'On mapping, you have got* [ROAD]...'
Officer A:	*'I'm by* [LOCATION], *is that right?'*
Officer B:	*'No, it's further round, near the church...do a left there'.*
Officer C:	*'[OFFICER C] to 3-5'.*
Whiskey 3-5:	*'Go on'.*
Officer C:	*'Can you speak to the IP and see if a laptop's been stolen?'*
Whiskey 3-5:	*'Confirmed'.*
Officer C:	*'I've found a laptop cable...'*
Whiskey 3-5:	*'Does that give a direction of travel?'*
Officer C:	*'It goes to a dead end...'*
Whiskey 1:	*'Whiskey 1 to control?'*
Control:	*'Go ahead'.*
Whiskey 1:	*'Another property is open.* [OFFENDER] *may still be inside'.*
Whiskey 3-5:	'[Requests location of this address]'
Whiskey 1:	*'...outside IP's address, go back...second right...'*
Golf 3:	[Talks to control, should the dog be cancelled, as lots of officers have been running around the alleyways to the back of the property]

This extract shows part of the radio communications during the response to a *break-in in progress* (burglary), where several officers were already at the scene, searching for the offender, and other resources were en route. As can be seen, officers are using the talk group to directly communicate in order to co-ordinate their response, with the controller playing an ancillary rather than a leading role. Interestingly, although the sergeant involved provides some leadership to the other units – for example, directing units during the search – none of the units involved in the example is demonstrably *in charge* of co-ordinating the response. Instead, the units involved jointly make sense of and determine the response to the incident, based on the working hypobook (break-in in progress) and the situation as they find it. This also shows that the controller has to repeat the incident details several times, either because a new unit has become involved (dog handler) or because the details have been forgotten (Whiskey 2).

7.5 Closing the Incident

Once the incident has been resolved, the officer will radio the controller with a final update that summarises their assessment of the incident and the actions that were taken. This narrative could be as short as *'One under arrest for being drunk and disorderly – transporting to custody'* but may be more lengthy for complex incidents. The controller will add this final update to the incident log, which is then closed. As soon as is practicable – which may be several hours later, if an arrest has been made – the officer will update the front of his or her pocket notebook with his or her formal account of the incident. This account may go on to form the basis of several items of crime file paperwork, as well as supporting the officer's recollection of the incident during any future court appearances.

Now that the incident has been resolved, the officer is able to formulate his or her impressions regarding the events and can reorder the fragments of information from the back of his or her notebook into a narrative of the incident in the front. This narrative coherently relates what he or she saw, the decisions and actions that he or she took and the outcome. His or her entry in the front of their notebook is therefore not merely the relaying of a series of events but also involves a retrospective interpretation of the meaning of those events, i.e. sensemaking. The representation of incident information from the back of the notebook to the front should not cause concerns regarding evidential accuracy, as no information has been lost; however, it does reveal that the notebook serves an additional sensemaking function, beyond merely being a personal record of events. The officer's retrospective narrative of the incident is the bridge between the disjointed *raw* information (captured informally in the back of the notebook) and the formal sequential record of events in the front. In this way, officers' pocket notebooks perform a similar role to the call handler's notepad and the IMS. However, the pocket notebook has a much lower potential to support collaborative sensemaking during the incident, as unlike the IMS, its contents are not readily accessible by other agents.

8
DISTRIBUTED COGNITION IN ROUTINE INCIDENTS

In this chapter, the process of incident response is related to the concept of distributed cognition. The examples are considered in terms of the types of artefact or collaboration that actors have available and how these can influence sensemaking.

8.1 Introduction

Table 8.1 summarises the routine incident response process composed of three high-level tasks, each of which is performed by specific agents from across the C2 network. These agents make use of several artefacts that support them in completing their tasks and mediate their communications with one another.

Figure 8.1 gives a view of the call-handling process as a sensemaking activity, showing the parallel iterative sensemaking processes that take place during the call and the important role that artefacts play in supporting this activity. The call handler's notepad acts as a private cognitive artefact that assists the call handler to engage in a frame-seeking activity. It does this by functioning as a shoebox for the temporary capture of potentially relevant details during the initial questioning and the verification process with the caller. During creation of an incident log, the IMS acts as a resource for action, as it requires the call handler to grade the call and select from a defined set of formal incident types to classify the nature of the incident. Determining the most appropriate incident type is not a matter of choosing from a list; instead, the call handler comes to recognise the type of incident that the incident represents during the question-and-answer session. Once identified, the incident type then acts as a frame that cues further data collection from the caller by tailoring the call handler's questions. In this way,

Table 8.1 Key Incident Response Tasks, Personnel, Their Locations and Associated Artefacts

TASK	ROLE	LOCATION	ARTEFACTS
1. Take 999 call	Call handler	Force Communications Centre (FCC)	Call handler's notepad Incident Management System (IMS)
2. Support responding units	Controllers	21 operational command unit (OCU) control rooms (local controller) FCC (traffic controller)	IMS Digital radio
3. Respond to incident	Responding officers	Patrolling 21 OCUs (local units) Patrolling 3 regional areas (traffic units)	Digital radio Pocket notebook

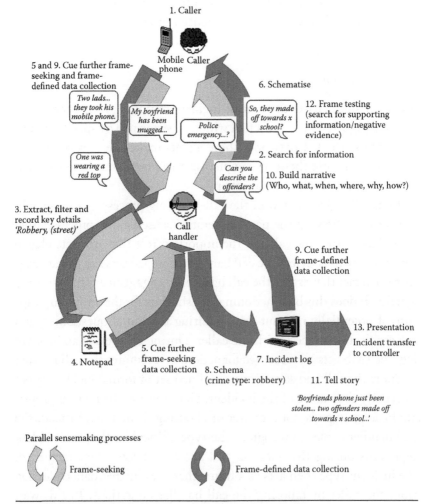

Figure 8.1 Representation of the call handler's sensemaking process.

once the appropriate frame has been identified, the decision of how to manage the call becomes a self-fulfilling one.

The use of formalised language in log entries helps to clearly and unambiguously summarise the incident to the controller. At the same time, the inability to edit entries encourages call handlers to be clear on the nature of the incident before they begin to type. This may serve the function of the incident log as an auditable record of the incident response, but it does not support the iterative nature of the sensemaking process and, along with the inflexibility of the log structure, is likely to be a driving force behind the requirement for a paper notebook *shoebox* to support the initial problem-framing process.

In this manner, the artefacts help the call handler to construct a succinct narrative to make sense of the incident, which they use to translate details from the unstructured account (caller) to a formal record (IMS) via an informal temporary store (notepad). Whilst the use of the notepad leads to a small amount of duplication of activity (i.e. capture of information twice), it arguably saves time overall, through the production of a clear, concise summary of the incident.

8.2 Allocating Resources to Incidents

The role of initiating the police response to a new incident is performed by the controller, who assesses the nature and priority of the incident, selects appropriate resources to respond and then provides support to them during the course of the incident. Figure 8.2 shows the controller using the IMS open incidents list (**1**) and a copy of the duty sheet (**2**) to manage the resourcing of new incidents as they come in. The right-hand screen is for the radio touch-screen software (**3**), whilst the left-hand screen (**4**) is shared between the GIS* (shown), the Police National Computer (PNC) application, email and intelligence systems. The controller also has some notepaper (**5**).

Controllers are responsible for managing the allocation of not only the immediate response incidents, but all open incidents in their area. Figure 8.3 shows part of an open incident list from Warwickshire Police. Each incident is given a unique reference number, the last four

---

* Geographic Information System.

Figure 8.2 Controller's workstation.

Incident	Location	Call Type	Att. Time/Task Desc.	Priority	Status	Resource
0282		CONCERN	CONCERN (NOT MISSING)	Priority	Received	
0254	STORES LTD.	MISPER MED	MISPER MEDIUM RISK	Priority		
0240		R.T.C.	A02	Scheduled	Arrived	A02(E)
0239		THEFT S	N02(E)	Priority	Arrived	N02(E)
0216		SEXUAL OFF		Scheduled		
0205		NEIGH DISP		Scheduled		
0186		FIREARMS	* *	Priority		
0178		NEIGH DISP	N39	Scheduled		
0160		DAMAGE	BE54	Scheduled		
0149		DAMAGE	N15(E)	Scheduled		
0145	BANK	DAMAGE	BT51	Scheduled		
0138		THEFT		Scheduled		
0095		VIOLENCE	84 16:00	Priority		
0088		TFMV	BE62	Res Wd		
0072		TFMV	NN63	Scheduled		
0071	EXACT ADDRESS NOT	DAMAGE		Priority		
0008	EXACT ADDRESS NOT	HATE INC	N14(E) RESULT	Priority	Assigned	N14(E)
0655		BURGLARY		Emergency	Cleared	N15(E)
0626		HATE INC	N13(E)	Priority	Cleared	N03(N)
0616		BURGLARY	AFTER 18:30	Scheduled		

Figure 8.3 Screenshot of controller's open incident list.

digits of which will reset at midnight. From Figure 8.3, it can be seen that Warwickshire Police dealt with over 650 incidents in a single day (across all five districts). There are only a limited number of resources available to deal with the large number of incidents, meaning that it is common for all units to be allocated to incidents at any one time, raising problems for the controller in allocating new incidents. The right-hand column in Figure 8.3 shows the resources that are currently listed as dealing with an incident and indicates that several logs are not currently allocated.

New incident logs are added to the incident list, and the controller is able to see the location, call type and priority even before they select the log. Opening a log automatically creates an entry (e.g. '*Incident Transfer Accepted by terminal D2R1*') and transfers the ownership of that incident to the controller. This update to the log is then visible to anyone else who has the log open, so the call handler is made aware that the controller now has the ownership of the log. The controller will first review the incident summary, before beginning the process of allocating resources to it.

When a new incident comes in, controllers should allocate the nearest available unit to it, referring to either the electronic (Warwickshire Police) or paper (WMP) duty list. However, during busy periods, controllers often struggle to keep track of the location and status of units. Officers will generally radio the controller when arriving at an incident, but only notify them of their location and status infrequently. This is because regular location updates are impractical over a busy radio network, and officers often visit multiple locations to make enquiries or return to the station to complete paperwork.

Whilst the GIS provides controllers with a map that shows the position of response units, it is infrequently used during incident management and is not drawn on when resourcing incidents. Because of not knowing the location or availability of units, controllers frequently make the radio announcement '*Any unit available for an immediate?*'. Busy officers will wait to hear if anyone else is able to attend before volunteering.

Control: '*Any unit available for an immediate...? There is a fight at* [LOCATION]...'
[No response]
Control: '[GIVES FURTHER DETAILS]'
[No response]
W3: '*Whiskey 3: we can divert from this arrest if there is nobody else available?*'
Control: '*Whiskey 3: Yes, I think you'll have to, as all other units are committed*'.

The GIS system shows the locations of officers and incidents, but it updates slowly and requires multiple page refreshes to zoom from a county-wide view, down to detailed views of towns and streets. This renders it unsuitable for quickly checking unit locations prior to allocating incidents. This is especially true for rural forces (such as Warwickshire), where some units are spread out across large areas, whilst others are densely concentrated in towns, requiring frequent changes to map scale in order to visualise the status of resources. Controllers therefore rely on the IMS open incident list as the primary resource management tool, with the GIS often hidden behind other, more frequently used applications – such as the PNC.

In order to utilise the IMS open incident list to manage multiple incidents, many controllers customise their use of the system. In Figure 8.3, the controller has modified the 'Att. Time/Task Desc'. (attendance time/task description) column to enable them to plan the activity during the course of the shift. Glancing at this column helps them keep track of what is happening and what actions they should take next: the entry of a call sign (e.g. *'A02'*) indicates that an incident has already been allocated to a unit to deal with later. The controller has made a number of entries to prompt themselves to allocate incidents at certain times – for example, *'B4 16:00'* and *'After 18:30' (which relate to the availability of the caller)*; they have also indicated those incidents that have not been allocated (with '…') and the unallocated 'Priority' firearms incident is given stars, to indicate its importance (i.e. "*…*"). Once an incident has been resolved and the corresponding log has been closed, it disappears from the open incident list. Thus, the IMS open incident list resembles a 'to do' list, both in form and function.

Whilst the IMS gives the address for an incident, this is in text, rather than map form, and the officer location information is absent. For controllers to be able to frame the incident management problem in spatio-temporal terms (i.e. *'which is the most appropriate unit to allocate this incident to?'*), they would also need to make reference to the GIS or radio units for status and location updates. With large numbers of incidents to allocate and a high volume of radio traffic to respond to, controllers concentrate instead on rapid allocation (*'Any unit available…?'*) over efficient allocation (i.e. closest available unit). Therefore, whilst the IMS interface design enables controllers to manage the resourcing of both urgent and non-emergency incidents, its use appears to encourage them to frame the problem of resource management in an overly simplistic manner. This assessment is supported by the observation that controllers making *any unit* requests frequently omit to give the location. This makes it hard for officers to determine whether they are close enough to be able to respond in a timely manner. Often, it is only once a unit has volunteered to attend that the controller divulges the location, at which point it may become apparent to another unit that they are closer to the incident.

Control: 'Sierra 3, can you go to [LOCATION]?'
S3:   'Yes'.
S2:   'Sierra 2 here, we are closer, so we can make'.
Control: 'OK, Sierra 2 please attend, Sierra 3 stand down'.

The high volume of radio traffic (irrelevant to them) means that individual officers only selectively attend to radio broadcasts, particularly when actively dealing with an incident. Additionally, officers do not have access to IMS or GIS whilst out on patrol. Consequently, officers are largely ignorant of each other's locations and status and so are unsure who is available or closest to a particular incident. They are also unaware of the list of unallocated incidents, limiting their ability to make sense of and co-ordinate their response to the wider workload problem.

The lack of data access for mobile officers, combined with the reliance of controllers on (a) the IMS to frame the resource management problem and (b) officers to volunteer for incidents means that neither are in a position to address the question of how best to allocate resources to incidents.

8.3 Making Sense with Artefacts

In making sense of the incident call, the call handlers are supported by their notepad and the IMS log, which act as external representations (Table 8.2). These artefacts perform many of the roles identified in Pirolli and Card's (2005) representation construction process as the call handler captures fragments of information from the caller, reinterprets these into a recognisable incident type and presents them as a coherent narrative summary that can be passed along the incident response process and dealt with. A similar process is in evidence at the scene of an incident, where officers make use of the back of their

Table 8.2 The Main Artefacts Involved in Police Incident Response Sensemaking

ARTEFACT	FORMAL PURPOSE	SENSEMAKING ROLE	EVALUATION FOR SENSEMAKING
Call handler's notepad	A temporary, unstructured record of key call details.	A private resource for action, cueing of frame-seeking activities (shoebox).	Pro(s): • Flexible – supports unstructured data capture. Con(s): • Required to re-enter information. • Low distribution potential.
IMS log	• The sharing of incident details. • Response initiation and risk analysis. • A permanent record of actions taken.	• A shared resource for action, a prompt for frame-defined data collection. • Capture and sharing of the formalised incident narrative.	Pro(s): • Enables rapid capture and dissemination of key incident information. Con(s): • Lack of access for response officers can create an information bottleneck. • Inflexible structure appears not to support frame seeking.
Digital radio	Enables incident response communication.	• Provides main means of communication between controllers and officers. • Enables officers' frame-defined data collection (via controller).	Pro(s): • Supports collaborative sensemaking. • Enables mutual monitoring within talk group. Con(s): • Use of point to point excludes other users and limits information exchange.
Pocket notebook	An officer's formal record of the incident and his or her actions.	• A private resource, cueing frame-defined data collection and a reframing of the incident. • Capture of formalised retrospective incident narrative.	Pro(s): • Supports unstructured data capture. Con(s): • Required to re-enter information. • Duplication of existing information (within IMS log and databases).

pocket notebooks, along with the IMS log and police databases (via the controller), in order to support the process of making enquiries. Once an incident has been resolved, the officers involved will (as soon as practicable) write up their account of the incident in the front of their notebook; this account may subsequently be *presented* as evidence within the crime file paperwork or during court proceedings.

These two iterative processes initially feature the use of informal, private cognitive artefacts (notepad, back of notebook), prior to the transition to formal, public ones (IMS, front of notebook). The reason for this appears to be that the unregulated nature of informal artefacts affords the flexibility to support frame-seeking, by allowing for rapid, unstructured capture and manipulation of information, which may go on to form part of the formal record or equally play no further part in the response (cf. Kirsh 2013). This echoes Baber et al. (2006) and Reddy et al.'s (2009) differentiation between the formal and informal artefacts that are used in sensemaking and reporting. Baber et al. (2006) describe how narratives are constructed to develop the crime scene investigation from informal sensemaking to formal reporting. In a similar manner, after an incident is closed, two narratives constitute the formal record of events. Firstly is the IMS log, which begins with a formalised account of the incident transformed from the caller's unstructured account, before giving an *in-the-moment* account of the incident response as a series of time-stamped event updates that reflect the twists and turns of the ongoing sensemaking process that took place during the incident. The second narrative is in the front of the pocket notebook, and comprises the officer's retrospective account of their thoughts and actions in relation to the incident and thereby comprises their ultimate frame for that incident (cf. Klein et al. 2006a). Both of these narratives fit the events of the incident into an established incident response framework that is recognised across the C2 network and wider judicial process, allowing them to be acted upon both during (IMS log) and long after (pocket notebook) the incident.

8.4 Making Sense through Artefacts

Sensemaking during routine incident responses is concerned with framing the problem; once an incident has been defined in terms of a recognisable *type*, standard operating procedures can be applied in order to guide the process of resolving it. The process of framing the problem is distributed across the individuals within the C2 system, and again, they are supported in this by the artefacts that are available to them. However, it appears that when dealing with emergencies, the various agents within the system interpret the sensemaking problem facing them differently, depending on their role.

As they make sense of emergency calls, call handlers appear to interpret the problem in terms of identifying the correct call *type*, which could be rephrased as the question, '*How can this incident be formalised within defined emergency response parameters?*' Interestingly, the caller appears to act as a resource for action that supports frame seeking, as each response from the caller prompts the call handler to ask a further question/seek clarification, until the call handler has established the nature of the incident (Figure 7.3). Once an appropriate incident type has been identified, this cues the call handler to engage in frame-defined data collection, further tailoring the questions that are asked of the caller to gather relevant details (Figure 7.2).

Controllers appear to view incident logs as multiple competing tasks to be allocated to response units, i.e. '*How does this incident fit within the wider service demand?*'. Controllers lack suitable artefacts to assist them in the task of identifying the most suitable resource to respond, which frequently results in their request for *any unit* to attend (Figure 7.3).

Responding units initially frame the incident in terms of assessing and preparing for anticipated risks, i.e. '*What do I expect to encounter at the scene?*' They are assisted in this by the IMS log – via the controller – though the controller's ability to support the responding officers (and thereby act as a resource for action) is often limited (Figure 7.5). The fact that controllers rarely offer supplementary incident information without it being requested by an officer also suggests that it is of no relevance to the controller's own sensemaking requirements as a *resource allocator* (Figure 7.5).

Once officers are on scene and have the situation under control, the sensemaking activity becomes one of establishing what has happened and what response is necessary, i.e. '*How do I resolve this incident?*' Response officers are supported in this task by the individuals whom they encountered, the scene itself, the controller (and their databases) and their own notebook – all of which act as resources for action, cueing officers to perform specific data gathering and frame-related activities (e.g. seeking, questioning and re-framing the incident – Figure 7.6).

8.5 Improvised Artefacts

Officers occasionally find themselves in situations where no suitable artefact exists, in which case they will improvise new ones. When an arrest is made, the prisoner is transported to the station, where the arresting officer will need to give an account to the custody sergeant, who will decide whether to approve the prisoner's detention. This account needs to include certain details, such as the time and location of the arrest, as well as the time of arrival at the police station. However, during busy periods (and particularly in large rural forces), it can often be over an hour from the time of arrest to speaking to the sergeant, during which there may be limited opportunity to make a pocket notebook entry. Officers regularly wear disposable gloves during searches and arrests, and, in a similar manner, to emergency medical practitioners (O'Connor 2010), they will often write on the back of the glove in situations where it is not practicable to make an entry in their pocket notebook.

Call handler: *'THERE ARE 3 LADS TRYING TO BREAK INTO CARS I AM WATCHING THEM ON CAMERA'.*
Call handler: *'CAR BEING BROKEN INTO NOW'*
Call handler: *'THERE ARE 3 IC1 YOUTHS WEARING DARK CAPS TROUSERS AND JACKETS – ONE HAS WHITE STRIPES DOWN TROUSERS AND WHITE TRAINERS'*
Call handler: *'CALLER SAID THAT THEY ARRIVED IN A CAR BUT HE DOESNT KNOW WHAT TYPE – CALLER IS LOOKING FOR IT ON HIS CAMERA'*

As the caller gives his or her initial account, the call handler is able to quickly note key details of the incident on his or her notepad, such as location, the type of incident and the persons who are involved. The call handler will then take control of the conversation, using the information on the notepad, firstly to cue further questions to and clarification from the caller and secondly to check his or her understanding by verbally summarising the incident back to the caller. This is an iterative process, with the call handler's notes and the caller's responses cueing further questions from the call handler, until they are clear as to the nature and severity of the incident. This process is taught to call handlers as *active listening*, i.e. '*...receiving information, clarifying, summarising and checking the message in order to reach proper understanding*' (Warwickshire Police 2005, p. 19).

The type of incident then defines the information that the call handler is required to collect.

Published entries cannot be amended or deleted; if any corrections are required, they must be added to subsequent lines of the log. Consequently, call handlers try to form a coherent incident summary before they begin to type and to enter information in concise statements.

8.6 Making Sense through Collaboration

During any of these stages, new information could prompt agents to question the frame and subsequently preserve, compare, re-frame or begin frame seeking again. Whilst these various re-evaluative sense-making strategies can be undertaken by individuals, it appears that collaboration plays an important role in the identification, investigation and resolution of inconsistent data and violated expectations. For example, two controllers, a call handler and multiple officers at the scene engage in a collaborative process of *bridging the gap* during an incident response. This begins when officers responding to a 'Break-in in progress on X Road' have their expectancies violated, cueing them to question elements of the frame with the controllers (via the radio). This prompts one of the controllers to engage in re-framing, first trying to ring the caller back (from IMS log details), then via the call handler (reviews call recording, changes location to 'X Street'). Further information from the second controller calls the alternative frame into question (invalid house number on IMS log), and the two frames are compared by the responding officers (attend both locations, check properties and speak to residents).

One of the controllers then engages in frame seeking (checks for the caller's details on the voters' database – negative result); the combined results of these investigations (no signs of break-in at either location; caller's details appear to be false) lead to the identification of a new frame of 'False call', and the response is abandoned. This description indicates the socially distributed nature of incident response sensemaking, with multiple agents working closely together to achieve a shared sensemaking aim. Interestingly, this is at variance to one of Perry's (2003) characteristics of socially distributed cognition – the notion that tasks must be organised such that they can be divided into components that can be performed by individuals, before being reintegrated again. There are no clear dividing lines between the actions of the various individuals, particularly as all of the activities contribute to the final frame that is used to describe the incident. Additionally, this activity has taken place on an *ad hoc* basis, with no single person in charge of co-ordinating the tasks to make sense of the incident. This sensemaking activity is supported by the artefacts that are used to mediate communications (IMS and radio) and others that act as resources for action, with the IMS log and police databases cueing frame-questioning and frame-seeking activity from the controllers.

The use of shared radio talk groups also enables agents within the C2 system not directly involved in an incident response to monitor events and (free airtime permitting) contribute to the sensemaking process. This enables the responding officers to benefit from one another's diverse experiences, which are not readily available through the formal incident records that are available in the IMS and which controllers are frequently too busy to interrogate. More infrequently, when complex emergencies involving multiple units take place, the responding units take control of the talk group and use it as an open forum for collaboratively making sense of the situation. For example, officers at the scene communicate directly with one another in order to elaborate the 'Break-in in progress' frame. However, the high levels of radio traffic preclude these forms of collaborative sensemaking from becoming a more frequent occurrence.

Within the control room environment, local controllers appear to engage in reciprocal monitoring, supported by their body of practice (Heath and Luff 2000) that enables them to closely co-ordinate their support activities and shared use of artefacts without the requirement for explicit communications.

8.7 Conclusions

Two important factors that dictate how incident response C2 performs as a collaborative entity are the social and organisational characteristics of the system. Organisational structures are designed with the purpose of the system in mind; social processes, on the other hand, evolve organically over time as people reorganise the information-processing environment (Perry 2003), are influenced by the organisation's cultural heritage and are highly resistant to interference (Hutchins 1995b). The design of the incident response C2 structure is intended to facilitate the efficient handling of emergency calls and co-ordination of resources to resolve multiple simultaneous incidents. This has been achieved through the specification of roles and functions within the network and the provision of technology to support the exchange of information – including the IMS log and digital radio network. Within this designed system, the focal point of sensemaking *product* from both call handlers and responding officers – the *centre* of the system – is the controller.

As this chapter describes, controllers are not in a position to co-ordinate responses, due to a lack of suitable artefacts and competing demands from their high workload. The lack of data access for mobile officers, combined with the reliance of controllers on (a) the IMS to frame the resource management problem and (b) officers to volunteer for incidents, means that neither are in a position to address the question of how best to allocate resources to incidents. Thus, there is a discrepancy between the centralised nature of the C2 network and the distributed nature of sensemaking activity, with the supporting technologies set up to facilitate the former (formal) arrangement rather than the latter (evolved) reality. This is reminiscent of Landgren's (2004) description of the tension between centralised incident response C2 and the responding units *preferential right of interpretation* of the situation, i.e. they are the only ones who have *eyes on* the situation and who are able to definitively determine what is going on. Landgren (2004) describes how this lack of direct access to information prevents responding units from being able to interrogate it and find inconsistencies – something that appears to be a key requirement for questioning and reinterpreting the frame that is used to respond to an incident.

In terms of the social processes that are involved in collaborative sensemaking, unsurprisingly, routine incident response C2 does appear to function as a community of practice, that is, as an established organization of individuals operating within a well-defined domain and that '...*share a common set of patterns of interpretation, implicit assumptions, and beliefs...*' (Burnett et al. 2004, p. 12). In addition to the stability of team membership, common training and experience and shared language, it appears that artefacts play an important role in supporting the effective realization of this community across a distributed network during fast-paced emergencies. Equally, it is likely that the artefacts in use are only able to act as frames in the way that is described in this chapter because of the community of practice, i.e. the highly compact, formalised nature of communications used in the IMS logs and radio transmissions frequently require the recipient to be in possession of detailed knowledge of the domain in order to be able to understand them. Where questioning of a frame does occur, any further comparison, reframing or new frame seeking is still defined in terms of the established language and procedures, i.e. routine incident sensemaking is a culturally defined activity (Weick 1995).

9
Responding to Major Incidents

This chapter details the challenges that are associated with multi-agency sensemaking and co-ordination during major incidents. The chapter is based on a critical instance case study of the defence of Walham electricity substation from the rising floodwater on July 2007.

9.1 Introduction

In Chapter 8, sensemaking was considered in terms of single-service, *routine* incident response. We noted how the different actors (and the artefacts they use) work to turn unstructured information into the *sense* of the incident to which they need to respond. As the incident unfolds, new information becomes available, or situational factors change, or different actors provide different perspectives on the nature of the incident or the suitability of the response. This means that the sense of the incident can change during the response, and it is important to manage the response accordingly. Within a single organisation, the manner in which the response can be managed adaptively is partly a function of the flexibility of the procedures that are followed and partly a function of the ability of the actors who are involved to develop a sufficient *common ground* to be able to appreciate the need to modify their response. In the incidents that were considered in Chapter 8, one can further assume that the overall *goal* that was being pursued was the effective resolution of the incident, where the understanding of what constituted an *effective resolution* could be understood in terms of police policies, procedures and processes. If one imagines an incident in which there could be competing resolutions to an incident, e.g. taking an offender into custody versus releasing the offender with a caution, and that the two parties involved in deciding on a response had strong views of one or another resolutions, then one could imagine some debate between the parties. Taking this example a little further, if one can imagine an incident in which there are more than one goal; more than one notion of what constitutes effective resolution; more than one set of policies, procedures and processes and more than one version of common ground, then one can appreciate the challenge that is posed by a major incident.

In this chapter, our concern is with the challenges facing the emergency services when trying to make sense of multi-agency, major incidents. Major incidents can be separated into the following four main phases:

1. Initial response
2. Consolidation
3. Recovery
4. Restoration of normality

Each phase involves a different set of activities and is consequently associated with different sensemaking requirements across the various levels of command. In broad terms, the *initial response* phase echoes the discussions of Chapter 8; an incident is defined for a specific emergency service and the response is mobilised. In general (in the United Kingdom), any member of the emergency services can declare a *major incident*. This raises some interesting issues relating to what might constitute *major*. For example, if the nearest accident and emergency unit in a hospital has four beds, then a major incident could be one that has more victims than can be handled by this hospital. In this case, the initial challenge of co-ordinating a response would be to ensure that all victims can be accommodated in hospitals in the vicinity of the incident. Alternatively, a major incident could involve significant threat to the public, such as a leak of hazardous gas, which might require cordoning off the area and issuing of instructions to the public to keep doors and windows closed. In both cases, there is a strong likelihood that the incident would be handled by a single agency with supporting roles being played by other agencies. This notion of *supporting roles* would probably be read with a sharp intake of breath by members of the emergency services because, if they are involved in what has been called a major incident, they will all play key roles in ensuring the effective resolution of the incident. This also implies that each agency, in terms of managing its own response, would seek to manage the incident in terms of its understanding of the situation and the activity that it is able to perform.

Whilst there are a number of common themes to routine emergencies and major incidents (i.e. urgency, risk to life and high levels of uncertainty), major incidents differ from routine emergencies in a number of ways:

- The scale of the problem (for example, the risk/impact may be to hundreds of lives)
- Complexity (both of the incident and the required response)
- Novelty (Major incidents represent an exceptional set of circumstances.)
- Timescale (The period of incident may last several days.)

Close cooperation between agencies is required in order to enable a coherent response to the incident. However, cooperation does not appear to come easily during crises; a review of the crisis response literature identified several problematical features that often characterise major incidents (derived from Dynes 1970; Auf der Heide 1989; de Marchi 1995; Crichton et al. 2000; Smith and Dowell 2000; Boin 2004; Boin and T' Hart 2007; Mendonça et al. 2007; Becerra-Fernandez et al. 2008; von Lubitz et al. 2008):

- Little or no notice
- Temporary, *ad hoc* teams that rarely (if ever) work together
- Multiple objectives and interdependent tasks
- High psychological demands, with people working under time pressure and in stressful conditions
- High levels of uncertainty (concerning both the nature of the problem and the best solution)
- Role specialisation, with the need to pool different types of expertise
- Improvised organisational structures
- Require the application of knowledge outside of traditional incident response domains.

The challenge of major incidents is therefore not only to make sense of a novel situation, but also to develop an appropriate response and to use non-standard organisational structures and procedures to co-ordinate the execution of that response. Given these difficulties, effective management of major emergencies would appear to be an impossible task (Boin and T' Hart 2003). Two recent disasters demonstrated a number of these features. During the response to the South Asian Tsunami in December 2004, local and international non-governmental organisations (NGOs) involved in the subsequent relief efforts were deemed to have failed to co-ordinate activity amongst themselves and with local government, engaged in competitive practices and displayed a lack of trust in one another (Bennett et al. 2006). In the aftermath of Hurricane Katrina in August 2005, the response was slow to mobilize, with the result that tens of thousands of survivors were left without resources for nearly five days (Schneider 2005; Chua et al. 2007; Kapucu 2008). Specific criticisms of the response included a failure to share information, poorly defined lines of command and a lack of trust between agencies involved in the response, widespread interoperability failures and a lack of awareness amongst response co-ordinators of the presence of agencies working on the ground (Chua et al. 2007; Rojek and Smith 2007). Case studies of earlier floods reveal similar evaluations; for example, Rahman (1996) found that there was a lack of co-ordination of the response to the 1988 floods in Bangladesh, including a lack of trust of NGOs by local administrators, resulting in their exclusion from planning programs.

This recurring failure of agencies to effectively co-ordinate their responses to emergencies warrants further investigation. About 11 problematical issues with multi-agency emergency response work have been identified from these earlier studies and are listed in Table 9.1.

Table 9.1 Key Issues in Multi-Agency Emergency Response

ISSUES IDENTIFIED FROM PREVIOUS MULTI-AGENCY EMERGENCY RESPONSE STUDIES	RESEARCH THEME
Response systems overwhelmed by the scale of the emergency	C2 structures
Poorly defined chains of command	
Slow mobilization of response	
Failure to share information between agencies	Inter-agency communications
Lack of awareness of the presence and activity of other agencies in the area	
Failure to communicate warnings and other information	
Lack of coordination between agencies	Common ground
Competitive practices	
Lack of trust between agencies and disagreement over who is in charge	
Interoperability failures	
Failure to fully integrate other agencies into the response	

The fact that these issues recur so often implies that they are inherent challenges that are associated with the command and control (C2) of multi-agency emergency responses.

9.2 Case Study: Walham Floods, 2007

This chapter presents a case study of the defence of the Walham Electricity Substation from floodwater during the 2007 *water emergency* in Gloucestershire. Despite challenging circumstances, this was a successful operation, mounted at short notice. However, the incident illustrates the difficulties that are faced in making sense of and responding to large, complex multi-agency emergencies. This incident has been explored in a previous paper (McMaster and Baber 2012), and this chapter uses this incident to explore the themes that are developed in this book. The activity description is based on the SME interviews, publicly available media and official reports of the incident. The interpretation draws on specific examples of the challenges that are faced by the emergency services as they tried to make sense of the incident. As in Chapter 8, the findings are discussed in relation to the three perspectives of sensemaking as distributed cognition (making sense with artefacts, making sense through artefacts and making sense through collaboration).

Chapter 8 described how sensemaking in police routine incident response appears to take place within a collaborative community of practice, where – despite often initially high levels of uncertainty around the nature an incident – established procedures can be applied by a stable network of agents who possess extensive common ground. In contrast, the characteristics of major incidents described in Section 9.1 (i.e. rare, complex and unique situations that require a large-scale combined response) would seem better suited to an exploration network. Therefore, the expectation was that sensemaking during the multi-agency major incident response in the case study would take the form of an exploration network, i.e. one that was formed on an *ad hoc* basis, where there was little common ground and where the goal was to develop new frameworks for interpreting the situation and guiding the incident response (Burnett et al. 2004).

9.2.1 Background: The Defence of Walham Electricity Substation

During flooding in July 2007, the Walham electricity substation was at risk of being inundated with water as river levels rose (Figure 9.1). The Walham substation forms part of the critical national infrastructure, supplying electricity to over 500,000 homes (an estimated 2 million people) in England and Wales (Snow and Manning 2007). If the site had flooded, then it was estimated that the electricity supply would be interrupted for up to three weeks (Gloucestershire Constabulary 2007).

On Sunday, July 22, a multi-agency operation was launched to prevent rising floodwater from overwhelming the Walham substation during the high tide that was expected during the night. (The section of the River Severn near to Walham is tidal.) The response involved hundreds of personnel from a number of organisations, including multiple fire and rescue services, the Environment Agency and initially personnel from several Royal Air Force (RAF) bases. The plan of action was to construct a series of flood defences around the critical substation switching room; this included the use of sandbag reinforcements, a one kilometre ring of the Environment Agency's modular flood barrier and deployment of specialist fire-and-rescue high-volume pumps to drain the site.

Despite the short notice and difficult working conditions, the various agencies were able to co-ordinate an effective response and prevented the floodwater from forcing the shutdown of the substation, buying time to construct semi-permanent flood defences around the site. In comparison, floodwater forced the nearby Castlemead

Figure 9.1 Floodwaters threaten the Walham electricity substation.

substation to shut down, cutting power to around 50,000 homes (Environment Agency 2007).

This incident took place during the wider water emergency that affected Gloucestershire in July 2007, which included widespread flooding, as well as the loss of drinking water supplies to much of the county. This 10-day crisis stretched the emergency services to the limit, forcing them to request assistance from the military. The incident at Walham represented a major incident in its own right and was a critical part of the countywide flood response. This case study is concerned with sensemaking at the bronze command level (i.e. at the scene) and between bronze and higher command levels during the consolidation phase at the Walham electricity substation.

9.3 C2 Structures

As was mentioned in Chapter 1, the gold, silver and bronze command structure ensures continuity between the strategic intent, tactical plans and operational application. During the operation to save the Walham substation, a number of alterations were made to the standard fire and rescue command structure in order to cope with unique features of the situation. These included drawing on resources from the Avon Fire and Rescue service under the *mutual aid* scheme. One such resource was the Incident bronze commander, who was therefore not directly part of the Gloucestershire major incident command structure. In response to this, the deputy chief fire officer (DCFO) from the Gloucestershire Fire and Rescue service was also deployed on-site, to act as *gold liaison* – a nonstandard role that was created for this situation. The DCFO had been identified during the Sunday morning strategic coordinating group (SCG) meeting as the best person to oversee the defence of Walham, having been told, '*...you go and save Walham, that's your job, get who you need to do it – we'll help you*' (Gloucestershire Chief Constable). The DCFO was *hands off*, i.e. did not play an active part in the command of the response, but provided input to Gloucestershire Fire and Rescue gold command on the progress of the response, acting as *eyes and ears* for the SCG. The DCFO also provided advice and support to the bronze commander when he experienced problems due to working in an unfamiliar county.

Changes were also made at the silver command level. In response to the protracted nature of the countywide emergency and the numbers of resources that are involved, the refuelling of fire and rescue appliances became a priority concern. Consequently, Gloucestershire Fire and Rescue modified their command structure by creating the role of *pseudo silver* – a command function that is dedicated solely to co-ordinating the refuelling operation. The overall fire and rescue command structure, in relation to the defence of the Walham substation, is shown in Figure 9.2, with the lines of communication that are shown by the arrows.

It is clear that circumstances required the fire and rescue service to modify their organisational structures away from the standard major incident arrangement; however, this led to problems in maintaining a consistent understanding across the command network. The critical role of the Walham substation was recognised at both the gold and bronze command levels, the bronze commander having been told, *'We have got to save this if we possibly can'*. However, at the silver command level, this view seems to have been somewhat hidden amongst

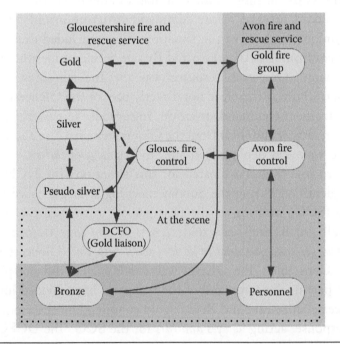

Figure 9.2 Bronze commander's view of the adapted fire and rescue C2 structure. (Dashed lines indicate probable lines of communication that are not observed by the incident commander.)

other competing priorities – at least in terms of the prioritisation of refuelling requirements by pseudo silver. It would appear that pseudo silver command considered Walham to be *one of many incidents*, rather than *the top-priority incident*. This is likely due to a combination of the following:

- The command *short circuit* created by having a direct link from the incident site to gold command (in the form of the DCFO)
- The parallel major incident command structures of Gloucestershire and Avon Fire and Rescue services
- The overspecialisation of pseudo silver – concentrating on one aspect of the tactical picture

9.4 Keeping Track of the Changing Situation

During a major incident, the fire bronze commander would normally be supported in his or her role by the deployment of an incident command unit (ICU) – a mobile command centre that provides command support staff, information technology infrastructure and briefing aides, such as maps and whiteboards (Figure 9.3). Command support staff perform control duties, not only co-ordinating the actions of fire and rescue personnel on-site, but also liaising closely with the other agencies responding to an incident. An ICU was not deployed to Walham during the initial response phase, and by the time the

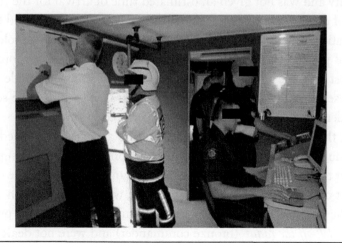

Figure 9.3 Interior of a fire and rescue ICU.

Avon Fire and rescue bronze commander took charge, there was no room on-site to deploy one. As a result, the level of command support available was severely limited, forcing the commander to run the incident with pen and paper as the only artefact support.

The bronze commander reported that the lack of an ICU increased his workload, in terms of the inability to use support staff and equipment to collect and represent the various elements of the situation and the response plan (such as floodwater depths). The bronze commander stated that he was gathering information, collating and assessing it and then formulating the response plan apparently almost entirely in his head (although this claim is debated at the end of this section). Thus, it was not possible for others to appraise themselves of the state of the incident and the response plan as they would normally be able to do from the status boards that are mounted on the ICU. Because of this difficulty in sharing information and delegating control tasks, the bronze commander was kept busy making decisions and giving orders to fire service personnel. This caused a C2 bottleneck that contributed to the communication difficulties with the Environment Agency, who described trying to get to speak to the bronze commander as *'like waiting in school queue'* (EA team leader).

The incident commander on the scene at Walham reported that when he made requests for fuel to be sent to the site (in order to protect *critical national infrastructure*), he was told that other incidents took priority and was not given an estimated time of arrival for the fuel. A lack of diesel for the pump generators had the potential to lead to the substation flooding, which would cause a wide-scale and prolonged loss of power. To prevent this, the bronze commander was forced to request that the DCFO (acting as *gold liaison*) contact pseudo silver and use his authority within Gloucestershire Fire and Rescue service in order to ensure that fuel would be delivered in time.

At the scene, the number of responding agencies, combined with communications problems and a lack of on-site command support, meant that the scale and pace of events began to overwhelm the *control* aspect of the bronze C2 capability. There were problems with tracking the progress of activities, with the result that some lower priority tasks *'fell off the radar'* (bronze commander) and were not dealt with, including clearing the non-essential vehicles parked on the approach road, which was to contribute to subsequent access difficulties that are experienced by the Environment Agency vehicles.

The fire and rescue service were responsible for the welfare of all personnel working on-site at Walham. They would normally manage site safety by manually logging everyone entering the *hazard zone* and where/what they are working on, for example, by using an entry control board (Figure 9.4). However, at Walham, there were too many people (over 100) moving on and off the site for them to be able to do this; instead, they were forced to look at the overall site safety.

The inability to monitor individual safety meant that safety was instead managed at the site level. The Royal National Lifeboat Institution (RNLI) boat crews were used to monitor water levels, as

Figure 9.4 An annotated fire and rescue entry control board. (Courtesy of Flickr.com, A Fire Entry Control Board with tags. Available at http://www.london-fire.gov.uk/. © Newswire.)

well as the welfare of personnel on-site and compliance with personal protective equipment (PPE; where it was available). Due to the lack of compatible communications between the different services, it was not possible to rely on radios to transmit the evacuation signal in the event that the risk from the floodwater became too great. Additionally, the whistles normally used by firefighters would not be heard over the noise of the pumps. An improvised evacuation signal was developed; an emergency services vehicle that was parked in a prominent position was nominated, and personnel were told that the evacuation signal was the use of the lights and siren on that vehicle.

Whilst the bronze commander stated that he managed the incident from a response plan in his head, this raises the question of just *how* was he able to do this, as it would not only involve constructing a mental map of the scene, but also a timeline of key events (including critical dependencies and key decision points) and a list of the available resources, their locations and activities. This notion also goes against the concept of a distributed cognition system. Rather than remembering everything at once, it is more likely that he was cycling through a process of information gathering, assessment, planning and execution, i.e. the incident command model (Chapter 5). Further, whilst there was a lack of standard command support artefacts, the bronze commander still had a number of resources for action available to support sensemaking activity. The concept of cognitive artefacts does not merely constitute *writing on things*; artefacts do not need to be designed or specially modified in order to act as representations – objects in the world can be used to represent information just because people decide that they do, for example, a knotted handkerchief (Vygotsky 1978; Norman 1993). Equally, people within the environment may be used to represent information and cue action (Hutchins 1995b). This was demonstrated in Chapter 4, which described how statements from the caller prompted the call handler to gather further information through supplementary questions, as well as the monitoring processes that are used by distributed (response officers) and physically proximal (controllers) agents within the system to support collaborative sensemaking.

Thus, whilst it may have felt to the bronze commander as though everything was in his head, it is the researcher's assertion that the bronze commander was able to draw on the environment and other people to act as resources for action to support the sensemaking activity.

In support of this process, the bronze commander walked round the site throughout the incident and (it is argued) during this was drawing on the environment in front of him as a representation to support his sensemaking framework, by attributing meaning to people and the environment. Thus, rather than trying to remember data (e.g. quantities of diesel, barrier construction progress), the commander would instead be able to remember the person that knows this information, or the physical location that represents the information and would then be periodically prompted to gather this information during his rounds of the site. For example, the bronze commander maintained close contact with an RNLI representative, enabling him to delegate the responsibility for monitoring water levels, welfare and PPE compliance; the bronze commander then only needed to remember that these tasks had been delegated and would be prompted to check by the presence of the RNLI representative, rather than trying to remember water depths or other data. Another example is provided by the need to monitor diesel levels for the high-volume pumps; the bronze commander designated a fire and rescue sector commander specifically for the high-volume pumps, who kept him informed as to the rate of diesel consumption, ensuring that this remained a high-priority activity.

This interpretation suggests that it would be harder to remember to keep track of issues that were not represented within the bronze commander's environment – either by physical objects or people. One such issue was the management of traffic outside the electricity substation, which was outside the bronze commander's environment and (arguably) was not represented by any physical objects or people. Consequently, the parked vehicles mentioned in Section 9.4 were not cleared, causing knock-on problems for the Environment Agency.

9.5 Conclusions

This case study provides an overview of the consolidation phase of a multi-agency major incident, showing how sensemaking was challenged by a number of factors. Retrospective interviews with personnel involved in commanding the response were used to overcome the difficulties in researching this aspect of incident response work. The case study reflects many of the challenges that are associated with major incidents that were described in Chapter 5, indicating that although every major incident involves a unique set of circumstances, the common themes that they exhibit may allow for generalization of learning points. Whilst the response to this crisis was rightly hailed as a success, the features of current incident response C2 networks, procedures and supporting technologies create additional barriers for emergency services personnel to overcome in order to make sense of the problem that they are dealing with.

10
DISTRIBUTED COGNITION IN MAJOR INCIDENTS

In this chapter, we consider ways in which *common ground* might be established and developed during the management of major incidents. Using the technique of the critical decision method, we compare and contrast the perspectives that are held by the different agencies that are involved in the response to the incident that is outlined in Chapter 9. This shapes the subsequent discussion on how sensemaking can be performed through the use of artefacts and through collaboration.

10.1 Introduction

The flooding incident at the Walham substation (discussed in Chapter 9) featured multiple agencies working on a shared task at the same location (i.e. the construction of flood defences and the drainage of floodwater from the site). By the consolidation phase of an incident, the responding agencies would (in theory) normally be involved in distinct tasks. During the defence of Walham, there was therefore a requirement for agencies to closely co-ordinate their activities to prevent them from working at cross-purposes.

The different agencies operated their own communications equipment: the Environment Agency (EA) and the fire and rescue service had incompatible radio systems, whilst the military did not have any communications equipment of their own and were forced to rely on the fire and rescue service to pass messages across the site for them. As a result, inter-agency co-operation at the bronze command level required physical proximity; this was not easy to achieve, given that the various agencies were spread around the site, and movement was restricted by floodwater, electrical hazards, construction activity and the movement of heavy machinery.

Co-ordination between the responding agencies appears to have been variable, with examples of both effective co-ordination and points of difficulty. The bronze commander was primarily concerned with the safety of personnel working in the substation; he therefore closely liaised with National Grid and Royal National Lifeboat Institution (RNLI) personnel (which required physical proximity), who helped to contribute to the risk assessment by setting safe working parameters and monitoring hazards. This collaborative process would likely have been made harder by the lack of artefacts and staff to support the bronze commander, combined with the associated high workload due to the requirement to monitor delegated *control* tasks. The difficulties associated with involving other agencies in the risk assessment process would have limited their ability to develop a full understanding of events and meant that they were unable to challenge the frame that had been used to make sense of the situation. For example, there was a delay whilst the Fire and Rescue Service found suitable vehicles to transport the military personnel onto the site through '...*100 metres of thigh deep, fast flowing water between the control point and the substation*' (military liaison officer [LO]). The EA teams (who at this time were not closely liaising with the fire and rescue service) were surprised to see this, as their own assessment was that the water was safe to walk through: '*The water was running over the road, but it was not deep or dangerous. The road was tarmacked and level and the water was only knee deep. The military probably thought it was deeper than it was*' (EA team leader). This vignette therefore demonstrates not only why different agencies are involved in major incidents but also why they need to collaborate to fully understand what is going on, which, in turn, requires the means to represent and share their interpretations across geographically distributed networks.

In order to determine whether it was safe for personnel to work on-site, information on a number of factors was collected and combined to produce an overall risk assessment for the site. The National Grid established safe working practices for personnel operating in *live* areas and defined a maximum depth for floodwater to reach before it would become too dangerous to remain on site. RNLI crews monitored water depths around the site and assessed floodwater risk to personnel, as well as reporting on compliance with personal protective equipment (PPE) use by personnel on site. The fire and rescue service took information from all sources, and from this, the bronze commander assessed the overall risk to personnel working on the site.

Risk factors were regularly checked, and the assessment of the level of risk to personnel working on the site was regularly reviewed. Near high tide, there was a concern that a breach in the flood defences could allow floodwater to suddenly overwhelm the substation, thereby causing an accident involving many people. This changed the perceived level of risk to personnel working on the site, and so the bronze commander took the decision to pull all but a few essential personnel out of the site.

The fire and rescue service and EAs' lack of mutual awareness of each other's roles, methods, processes and requirements generated logistical difficulties that delayed the response. This initial lack of awareness stems from the fact that these agencies would never normally work together and so were largely unknown to one another. The EA reported that their teams do not train with any other agencies; as a result, they were not used to having to collaborate and were unfamiliar with the major incident protocols. In this instance, the apparent simplicity of the problem, physical barriers to communication and high-workload demands on the bronze commander appear to have discouraged the type of collaborative inter-agency discussions that might otherwise be expected during a major incident. After the difficulties described above, the bronze commander and EA team leader realised that they needed to co-operate more closely and endeavoured to do so. The bronze commander commented that the fire service would normally seek to discuss the situation with partner organisations and seek consensus, prior to initiating the incident response. However, in this situation, he felt that there simply was no time to do this, and the constant requirement for command decisions meant that the briefings and situation updates to other organisations were limited.

The National Grid safety personnel advised the responding agencies at Walham on safe working practices; these restricted the use of lifting equipment in parts of the site, meaning that sections of the EA flood barrier would have to be moved into place by hand. It became clear to the EA team that they did not have enough personnel to complete the construction of their barrier in the time that was available. The military teams sent to the site were tasked with moving barrier components into place and assisting the EA with construction work.

The incident commander kept LOs from the military and RNLI close by, as he needed to maintain constant contact with these organisations. The fire commander initially thought that the EA were happy to be left to get on with their tasks, leaving him to focus on other aspects of the response.

The substation was surrounded by floodwater, and there was only one single-track road in to the site. The large articulated vehicles bringing in EA equipment had been held up in the queue of traffic outside the incident cordon, and the notification of their arrival was not passed to the EA team.

Once the EA team leader realised what had happened, he tried to get the lorries into the queue of traffic entering the site, but this was initially refused, as their size meant that all other traffic would have to be stopped to let them in and out. This delay put the completion of the barrier before high tide at risk. The EA team leader then approached the fire commander, and they discussed the problem, agreeing that the priority was the construction of the barrier. All other work and site traffic were stopped to allow the EA lorries into the site to be unloaded.

The military involvement in the Gloucestershire floods featured the use of LOs, including one on-site at the Walham substation for the duration of the incident. The role of the LOs was to function as an interface, bridging the gap between command and control (C2) networks, as well as the different languages, practices and perspectives on an incident. The purpose of this was to understand the requirement that the military were there to support, to provide the necessary assistance and to send updates back through their military command network. For example, on Friday, July 20, the joint regional liaison officer from 43 (Wessex) Brigade – having heard about the flooding problems in Gloucestershire and the use of military search and rescue* – decided to travel to gold control *to fight for information*, to establish whether there was any likelihood that large-scale military involvement might

---

* Military assistance for search and rescue is covered by a different legislative provision that allows for short-term urgent assistance of a limited scale.

be required. Consequently, by the time that military assistance was formally requested on Sunday, July 22, the mobilization of 43 Brigade was well underway, having begun in anticipation of the request.

The military LO for Walham described how he stuck to the bronze commander *like a leech*, in order to keep abreast of incident developments. The LO briefed each new team of military personnel before they entered the site and provided situation reports back to the military contingent in gold command every 15 minutes. The value of LOs was demonstrated by their widespread and effective use by the military and the number of *ad hoc* liaison roles that were created within other organisations (cf. Figure 5.2), in order to address the particular needs of the incident and to ensure the continuity of purpose across organisations and levels of command.

10.2 Common Ground

On the face of it, all agencies shared the same overall goal, i.e. to prevent the substation from flooding. They were also clear on what needed to be done; the incident was regarded as simple by both the fire and rescue service and EA, with the bronze commander describing it as a *no-brainer*. However, the responses to the critical decision method (CDM) probes listed in Table 10.1 show that the responding organisations were actually working toward different goals. Table 10.1 implies that they had very different conceptual frameworks for the incident, likely due to their different roles, responsibilities and experience causing them to focus on specific aspects of the incident.

The fire and rescue bronze commander had responsibility for the inner (hazardous) cordon of a major incident site; as a result, he was concerned to assess all the risks to personnel within that area. In contrast, the EA deal with water hazards regularly and assess risk on an individual basis; they saw the only other hazard as being the overhead electrical equipment, but were happy to work within the restrictions of the National Grid and to evacuate the site if required to do so by the fire and rescue service. Similarly, the two agencies understood the necessary response actions very differently. The EA viewed the solution as simply the deployment of their barrier, which would prevent the floodwater level from rising within the site at high tide. The fire and rescue perception was that they needed not only to stop the high

Table 10.1 Responses to CDM Questions from the Various Organisations, in Relation to the Risk Assessment of Having Staff Working Inside the Electricity Substation

CDM QUESTION	FIRE AND RESCUE	EA	MILITARY
Goal specification: What was your overall goal?	Maintain safety of personnel working on site. Prevent the substation from flooding (through sandbagging, barrier, pumps).	Construction of the flood barrier before high tide.	Provide maximum support to the bronze commander.
Cue identification: What features were you looking at when you formulated your decision (site safety)?	Predicted time and height of floodwater at high tide. Hazard conditions (advice from RNLI, National Grid, reports from firefighters). Control measures. Improvised evacuation signals.	Dynamic risk assessment – safe to work on site. National Grid guidelines on safe working practices. Evacuation signal from fire and rescue.	The risk assessment of the Fire and Rescue Service. State of floodwater across approach road – determined that this necessitated vehicular transport on and off site.
Conceptual model: Are there any situations in which your decision would have turned out differently? Describe the nature of these situations.	Evacuated all non-essential personnel near high tide, as risk of water overwhelming the defences rose.	Understand effects of high water on the site.	Continuous review of decision by all parties, under the chairmanship of bronze commander.
Influence of uncertainty: At any stage, were you uncertain about the appropriateness of the decision?	Constant review of decision; risk to personnel set against priority of goal; measures taken to manage risks.	Staff experienced in working in water hazard, had constructed the barrier several times that year. Trusted the National Grid as experts.	Could see that bronze commander was hesitant about military commitment to an unpleasant task.
Situation awareness: What information did you have available to you at the time of the decision?	Hazard assessment from National Grid: maximum safe floodwater level. Water depth and hazard assessment from RNLI and fire and rescue services. Compliance with PPE. Time of high tide.	Safe working practices from National Grid, experience of EA personnel.	The risk assessment of the fire and rescue service.

tide but also to deal with rising groundwater (which was entering the site inside the barrier). Thus, the fire and rescue response also included sandbagging the main switch room (the most critical point), as well as the use of specialist high-volume pumps to reduce the water level inside the barrier.

Under the Civil Contingencies Act 2004, the fire and rescue service and EA are Category 1 responders (HM Government 2005a), and both are used to being *in charge* of their own operations. Whilst the EA recognised that the fire and rescue service were in control of the site, and that they were concerned with the safety of personnel working there, some of their comments indicate that they were not comfortable with the command situation and suggest that they may not have recognised the *primacy* of the fire and rescue service. For example, they described how the fire service *'took control of the site'* (EA team leader) and that the EA were *'outnumbered 50:1'* (EA team member).

The fire and rescue specialist high-volume pumps used at Walham were considered one of the key elements of the incident response by the fire and rescue service. These pumps came from across the country; each one was brought onto the site by its own support vehicle, which then left the site and were parked up along the verge of the access road. This subsequently limited the ability of the larger EA lorries to deliver their equipment. As the EA teams viewed this fire and rescue equipment as unnecessary and merely an obstruction to their own activities (*'It was us that did it'*. – EA team member), they constructed another explanation for why fire crews from all over the country had been brought in, seeing this as *'an opportunity to dust off their gear'* (EA team leader). There was considerable media interest in the defence of the Walham Substation, with numerous television crews attending the incident site to make live broadcasts. This was interpreted by the EA teams as the reason for the actions of the fire and rescue service, which, to them, appeared to be driven by the public relations possibilities of a high-profile incident. They interpreted the movement of fire crews and equipment within the site not as the placement of critical equipment but rather as merely a tactic to raise their profile with the media '...[the] *press loved fire brigade flashing lights...units seemed to be moved around just to be high profile...*' (EA team member). At the same time, they felt that the role of the EA was being played down, as they were told to move their own vehicles *'out of the way'* (EA team member). The contrasting perceptions of the incident are hinted at in Table 10.1, which indicates that the EA were not aware of factors of the incident that the bronze commander considered critical.

It is clear that in the absence of any common ground between the agencies and where there was only limited collaboration (at least at the start) and understanding of one another's activities, the EA team searched around for a plausible explanation for what they deemed to be unnecessary and disruptive activity. They appear to have generated the frame of *PR exercise* (EA team leader) from the key data anchors they inferred from the situation, as described in the preceding paragraph (i.e. heavy media presence, unnecessary fire and rescue equipment from multiple services, equipment movements intended to raise fire and rescue profile, marginalisation of EA). The EA personnel interviewed still maintained this view several months after the incident had taken place, showing considerable resentment towards the fire and rescue service for the way they felt they had been treated, which at the time would not have helped motivate them to collaborate during the incident response. However, it is important to note that it is not possible to know if this account represents the EA personnel's assessment at the time, or if this was developed in the intervening period (approximately four months) between the incident and the interview.

The physical challenges (floodwater and the size of site), the lack of command support, the tight deadline and the apparent simplicity of the incident, all precluded in-depth discussion of the incident. It was not until there was a problem that threatened the success of the response that the two organisations began to discuss the incident in detail. However, the very different interpretations – both of the problem and the roles of the respective agencies in the solution – indicate the requirement for agencies to jointly engage in framing the incident, especially where they have limited common ground. At the same time, establishing a common framework for multiple agencies to apply to an incident is likely to take time and effort, something which is not readily available during an incident.

The EA team had a specialist role in the response, namely, the deployment of their flood barrier equipment; this was a task that they were familiar with, having already used the barrier several times that year. From their perspective, the incident was straightforward, and they knew what had to be done. However, they felt that the fire and rescue service were slow to adapt to the pace and nature of the incident. The EA considered that the fire and rescue service were *in the way* during the early stages, delaying the arrival of their equipment. The fire service brought in a number of appliances to deal with the incident; in the opinion of the EA, this seemed to be far more equipment than was required, as they felt that the barrier defences were already dealing with the incident appropriately. To them, the fire and rescue response appeared to be driven by public relations opportunities rather than saving the site. The EA personnel spoken to asserted that it was their equipment, personnel and knowledge that had been crucial in the defence of Walham; they felt that this went unrecognised, both by the fire and rescue service and in media reports of the incident.

The fire and rescue bronze commander was the overall incident commander and therefore had responsibility for the co-ordination of the whole multi-agency response, as well as the safety of all personnel working on the site. Therefore, from the fire and rescue commander's perspective, whereas the problem was simple, the management of the incident was much more complex, with many factors to consider, including numerous hazards. The incident commander identified a number of equally critical aspects to the flood defences, of which the EA barrier was one part. Due to the rising groundwater, eight specialist high-volume pumps were brought in from fire and rescue services around the country, to keep the floodwater level down within the substation. The incident commander felt that all of the agencies involved in the response were focussed on the same goal, rather than thinking that their own agenda was more important.

This example provides further support for the view that the bronze commander was drawing on other agencies as resources for action, by delegating activities to them and then using them as cues to prompt his own sensemaking activity, supported by the incident command model as a heuristic. However, if the bronze commander was seeking to distribute *command*, *control* and cognition across the site, then it would have been preferable for this to have featured in the incident planning from the start. Instead, as is often the case what seems to have happened is that the response was improvised in order to fit the constraints of the situation – such as the lack of incident command unit (ICU) – and the most appropriate command structure was therefore only arrived at towards the end of the incident.

The relationship between the fire and rescue service and the military at Walham was different. The military goal was to support the fire and rescue service in whatever way that they could. From prior training with the emergency services, the military LO identified the bronze commander as the individual who is *in charge* of the incident response; he then set about understanding what the bronze commander wanted the military to do and liaised with the various military units to see if this was possible. Where decisions were required from the military, these were passed up to their superiors at gold command, for example, the decision to deploy military personnel on-site without any PPE. The military relied entirely on the bronze commander for their understanding of the incident.

10.3 Making Sense with Artefacts

In a similar manner to the control room–based processes described in Chapter 4 (and explored by Blandford and Wong 2004), the Gloucestershire Fire and Rescue Service gold and silver command levels (based within the Gloucestershire Tri-Service Emergency Centre) were able to make use of a range of artefacts (including the Incident Management System [IMS], geographical information system, status boards, radio and telephones) in order to support sensemaking during the countywide emergency. In contrast, without access to the ICU, the bronze commander at Walham was left with only pen and paper to support him in making sense of and co-ordinating the response to the incident. Chapter 4 highlighted the value of informal, private artefacts (including pen and paper) in supporting frame-seeking activity; however, once the appropriate frame has been selected, they appear less well suited to support the C2 activities that are involved in executing the response than formal, public artefacts. For example, Chapter 4 described how the IMS can function as a *to-do* list, tracking resources against tasks and reminding controllers of the outstanding tasks to perform – something that ICU staff (drawing on artefacts) would normally carry out during a major incident. This chapter proposes the argument that, without the support of an ICU, the bronze commander relied on people and the physical features of the environment to act as representations instead, and that by moving around the site and referring to the incident command model as a sensemaking heuristic, the bronze commander was able to draw on these representations in lieu of a formal artefact-based *picture* of the incident. What the bronze commander appeared to lack was specific artefacts to help keep track of issues and actions that are not readily attributed to the elements of the environment that are available to him or her (such as off-site traffic).

As Chapter 4 described, the emergency services' use of formal, public artefacts to capture, represent and communicate information means that they represent the incident frame within a recognisable structure, enabling individuals to rapidly apprise themselves of the incident and contribute to sensemaking (for example, elaborating or questioning the adopted frame). Were the ICU available, the process of developing and maintaining the incident picture (i.e. frame-seeking, elaboration and questioning) could have been supported by formal, shared artefacts (such as status boards, maps and the IMS), allowing for wider dissemination and involvement of other personnel. This would then have assisted the bronze commander's ability to brief staff and delegate many of the *control* tasks, improving the efficiency of the C2 system and freeing up time for *command* activities, such as collaboration with the other agencies. The knock-on effects of the lack of suitable artefacts to support sensemaking appear to have included hampering framing the problem (Section 9.4) and collaborative sensemaking (Section 9.3).

10.4 Making Sense through Artefacts

Chapter 4 describes routine incident sensemaking as concerned with framing the problem. This is a distributed process, involving several agents from across the C2 network who are engaged in co-ordinated activity, supported by several artefacts. Within this process, the agents adopt clearly defined roles, and each predominantly tackles a discrete element of the sensemaking process. Once a routine incident has been defined in terms of a recognisable incident type, the response involves implementing SOPs. Similarly, sensemaking during major incidents also appears to be concerned with framing the problem. During the

Walham incident, the responding agencies framed the situation using existing incident types (flood barrier construction, flood response management – Section 9.2), which then cued the SOPs that they applied in response (i.e. barrier construction, sandbag defences, use of pumps to drain water), which could be distributed amongst the agencies that are involved.

10.5 Making Sense through Collaboration

Chapter 4 identified the social and organisational characteristics of the C2 network as having an important role in shaping routine incident collaborative sensemaking. These characteristics have again been identified in relation to both intra- and inter-agency collaborative sensemaking during major incident responses and are discussed, in turn, in this section.

10.5.1 Organisational Structures

The major incident command hierarchy described in Chapter 1 is designed to ensure the continuity of purpose across the command levels, providing mission command downward and oversight and feedback on progress upward. From a sensemaking perspective, three layers of command employ an iterative process to communicate and collaboratively develop their understanding of (i.e. make sense of) the incident and associated requirements. As such, mission command could be viewed as a means of developing and communicating the incident command framework. The incident response at Walham (Chapter 9) necessitated a number of *ad hoc* changes to the fire and rescue C2 network, for example, in order to provide workarounds to the lack of interoperability between fire and rescue services. Whilst necessary, these alterations resulted in confused lines of communication and undermined the command hierarchy.

Multi-agency major incident doctrine states that the emergency (and other) services should collaborate at every command level but maintain separate C2 networks (LESLP 2007; NPIA 2009). However, there appears to be a tension between the maintenance of single-service C2 networks that are designed to facilitate intra-agency activities and the need for collaborative sensemaking in order to correctly frame and thereby respond appropriately to multi-agency major incidents. This arrangement seems to have affected the social processes that are involved in collaborative sensemaking during the Walham incident.

10.5.2 Social Processes and Collaborative Sensemaking

The incident outlined in Chapter 9 shows the importance of collaboration in making sense of this incident, thereby demonstrating the insufficiency of the *individual-as-sensemaker* position for describing multi-agency major incidents. By definition, multi-agency emergencies involve multiple C2 networks and commanders, each of which is in possession of specialist knowledge and expertise that forms *the pieces of the puzzle* that are required to frame the problem.

Collaborative sensemaking during routine emergencies appears to take place within a community of practice, where – despite often initially high levels of uncertainty around the nature of an incident – established procedures can be applied by a stable network of agents who possess extensive common ground. In contrast, the multi-agency response to the case study in this chapter shares similarities with Burnett et al.'s (2004) exploration network, including the following:

- Networks that form and re-form depending on task and need
- Often poorly defined areas of common interest
- Short-lived, dynamic associations

The purpose of the exploration network is to exploit the breadth of knowledge of the diverse agencies involved, in order to develop innovative approaches that overcome the failures of existing interpretations (Burnett et al. 2004; Umapathy 2010). However, from this case study, it appears that whilst the major incident response network *could* function as an exploration network, the responding agencies (at least initially) tried to function as communities of practice, drawing on standard frameworks and procedures to understand and respond to the incident. On reflection, this is not entirely surprising, given the current major incident doctrine and the separate C2 structures. Two additional factors that were likely to have impeded the early development of an exploration network at Walham include (1) the lack of collaborative artefacts and (2) the effortful nature of collaborative sensemaking.

The lack of compatible communications technology at Walham means that physical proximity was required for collaboration, which was difficult on a large, flooded construction site with multiple distributed areas of activity. The inability to widely share representations of incident information severely limits collaboration during what can often be fast-paced events. This also curtails the ability for agencies to monitor one another's actions and proactively contribute, either to the sensemaking process or to deconflict proximal activities.

In routine incidents, artefacts are able to function as incident frames because of the community of practice, which enables the use of highly compact, formalised communications that require the recipient to be in possession of detailed domain knowledge. However, it has been argued that for exploration networks to function effectively, they require agents to make explicit that which is implicit, i.e. the various agencies need to articulate their interpretations of the incident in plain language, in order to foster debate and the formulation of new interpretations of what is going on (Burnett et al. 2004; Weick 2005). Consequently, whilst the support of ICU staff and formal fire and rescue artefacts would have undoubtedly assisted the bronze commander, the lack of common ground between the responding agencies suggests that it is questionable whether the presence of status boards and other ICU aides would have by themselves been of significant benefit in supporting inter-agency collaborative sensemaking.

The military and fire and rescue LOs involved in the incident response functioned to fill the gap within and between C2 networks that is caused by the lack of multi-agency technological interoperability. In sensemaking terms, the LO roles described in this chapter appear to share some similarities with response officers in Chapter 4, in that they arrive on-scene, try to establish what is going on (in terms of their own organisational drivers) and then report this assessment back in the language of their own organisation. Therefore, the LOs provide an important role in translating one organisation's sensemaking output into terms that are meaningful for another, which would not have been replicated merely by providing access to shared artefacts. However, the role of LOs as sensemaking *translators* falls short of exploration network–style collaborative sensemaking between the two organisations (i.e. to form new interpretations of the incident).

According to Weick (1995), collaborative sensemaking is a culturally defined activity. Chapter 4 demonstrated that within-service frame-seeking and modification are defined in terms of an established, formal *compact* language, underpinned by implicit assumptions – all of which presents a barrier to meaningful inter-agency interaction. Further, within exploration networks, it would be less obvious what information should be shared and with whom (Baber et al. 2008). In the absence of an obvious incentive to expend effort in collaborating on an apparently simple incident (cf. Chapter 8), and in the face of practical difficulties in doing so (lack of artefacts, overworked bronze commander), it is unsurprising that inter-agency collaborative sensemaking was initially limited.

10.6 Conclusions

The incident command model describes the approach that fire and rescue commanders adopt at major incidents, and whilst they would normally employ support staff to gather and record information and communicate plans, on this occasion, the absence of the ICU meant that physical objects and people were used as resources for action instead. According to the fire and rescue incident command model,[*] other agencies are seen as a source of information to draw on during response planning, and fire and rescue personnel represent resources to control to ensure that the plan is a success. Similarly, this chapter argues that the bronze commander drew on people by delegating tasks and then using them to cue his sensemaking activities, as is indicated by his responses to the CDM questions. Despite this argued for use of people and objects as artefacts, the lack of ICU is expected to have hampered the bronze commander's ability to brief personnel and delegate C2 tasks, thus increasing his workload.

This case study suggests one reason why major incident response co-ordination has proved to be a recurring problem, which is that they can appear deceptively simple. Organisations will recognise familiar elements of the incident that relate to their training and procedures, without picking up on the important characteristics of the situation that do not form part of their repertoire of past experience and are not

---

[*] *The Incident Command Model*, Avon Fire and Rescue Service presentation.

readily described within their formalised lexicon. As such, these characteristics could be considered to represent *unknown unknowns*, i.e. the organisations are unaware that they are missing important pieces of the puzzle. This major incident was described as being a *no-brainer* (bronze commander), and both the EA and fire and rescue service framed the problem differently, based on their training and experience and ruling out more in-depth collaboration with one another as there was not enough time to do this. This problem is not helped by the view that is prevalent within incident response doctrine, that the incident response should be undertaken using existing procedures (cf. LESLP 2007), thereby reinforcing the view that there is nothing fundamentally different about multi-agency major incidents.

This chapter demonstrates the fallacy within major incident response doctrine, of clearly delineated roles and areas of responsibility between the emergency services. For example, the fire and rescue service are to control the inner cordon surrounding the incident site, restrict access and oversee all the activities that take place within this area, whilst the police manage traffic, rendezvous points and other arrangements within the outer cordon. However, at Walham, the police were not in attendance (due to the countywide flooding emergency), and the fire and rescue service were not the only agency operating within the inner cordon. And the combination of the lack of inter-agency communications interoperability and the large numbers of personnel meant that the fire and rescue service were not able to maintain effective site access control. Similarly, during many of the large-scale crises, the entire area becomes a hazard zone, where no single agency is in control or has the whole picture, and roles and responsibilities become blurred, suggesting a collaborative approach to sensemaking and response co-ordination. This raises the challenges of interoperability.

11

THE CHALLENGES OF INTEROPERABILITY

In this chapter, we consider the relationship between sensemaking and interoperability. This develops some of the themes that were introduced in Chapters 9 and 10 (where we considered a multi-agency response to incidents). The concept of interoperability could be viewed in terms of the technical challenge of ensuring that the equipment used by different agencies is able to operate seamlessly. From this perspective, interoperability could involve defining (and implementing) common communication protocols. Thus, *sense* could arise from all people having access to the same data in the same format at the same time. Whilst this might be necessary for sensemaking, it is clearly not sufficient. Two case studies involving bomb attacks (in London and in Boston) help illustrate the challenges of interoperability.

11.1 Introduction

In Chapter 1, we noted that interoperability was also a social phenomenon, in which the challenge was to ensure that the agencies were able to align the procedures that they were following (to deconflict operations) and to develop an agreed understanding of the situation that they were facing (to seek to resolve the same problem). As we saw in Chapter 9, when agencies are pursuing different *understandings* of the situation that do not necessarily accord with each other, then it is possible for the activity of one agency to conflict with, disrupt or otherwise interfere with the activity of another agency and for the different agencies to act is if they were solving different problems. From this perspective, interoperability is concerned with being able to work together. Drawing an analogy with the notion of common communication protocols, it could be beneficial for the agencies to have a common *command language* through which they can share their understanding of the situation. Even if agencies share a common language (which, like common communications, is necessary for sensemaking), this is still not sufficient to guarantee sensemaking. The ways in which each agency will respond to terminology is not only semantic (in terms of an agreed set of definitions of terms) but also pragmatic (in terms of the actions that these terms will elicit from each agency, which, in turn, changes the importance of the terms across agencies and potentially leads to variation in *sense* of a situation). In terms of sensemaking, the challenge is to allow individual agencies to develop their own interpretation of the situation and to allow (if appropriate) this understanding to be shared with other agencies. Before considering how this might be possible, and how useful and appropriate it might be, the next section considers what the term *interoperability* means.

11.2 Defining Interoperability

The UK Emergency Services has engaged in a programme of interoperability improvement. Informing this programme is a definition of interoperability as '...*the capability of organisations or discrete parts of the same organisation to exchange operational information and to use it to inform their decision making*' (ACPO 2009, p. 14). The essential elements of this definition are that it concerns a *capability* (in contrast to specific functions or activity) to *exchange operational information* (implying a measure of efficiency in terms of the activity of communications systems, and for these systems to focus on mission-relevant information) in order to *inform...decision making*. (Implying a measure of effectiveness will improve the command and control [C2] for the mission.)

The Interoperability Continuum (NPIA 2010) has been adopted by the UK Emergency Services as the model for their interoperability programme (Figure 11.1). The model was originally developed by the US Department for Homeland Security for co-ordinated incident response planning, and presents interoperability in terms of five elements: (1) governance, (2) standard operating procedures (SOPs), (3) technology, (4) training and exercises and (5) usage, which are

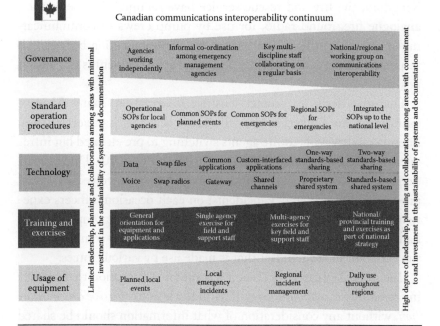

Figure 11.1 Interoperability Continuum.

measured across five levels of performance, ranging from the lowest *unacceptable* level (with limited leadership, planning and collaboration) to the highest *optimal* level (featuring a high degree of collaboration and co-operation).

11.2.1 Technology

Technology refers to the ability to exchange voice and data communications. The different emergency services exchange information relatively infrequently, as the majority of *routine emergencies* are single-agency incidents. The levels of inter-agency communication increase significantly for more complex or serious incidents; however, a lack of data exchange capability between services means that the control room communication is limited to telephone calls.

Agencies *could* have compatible digital radio communications systems. However, the different agencies might be free to purchase different models of handset, or have different service user agreements with the network provider, which means that not all functionality could be enabled for all personnel (ACPO 2009). In the United

Kingdom, the fire and rescue service have retained the use of their analogue fireground radio, for use by pump crews in communicating at the scene; this means that direct communication between all responding emergency services personnel at an incident is not always possible.

Due to a desire to protect sensitive information, many Airwave user groups are limited in the talkgroups that they can access; thus, interoperability will require agents switching to pre-defined but infrequently used interoperability talkgroups (ACPO 2010). During participant observation, one of the authors (RM) was involved in several pre-planned multi-agency incidents; on each occasion, officers experienced difficulty in finding the interoperability talkgroups on their handsets and had to be shown how to do it – further complicated by the variation in the handset that is made use by various forces.

The *technology* category reflects a common assumption that providing *the ability to communicate* will naturally lead to *effective communication*, without any consideration of what information should be shared between individuals and organisations and the manner in which this is to be done. The evident importance of technological interoperability as the foundation for effective communications has meant that it has tended to be the dominant focus of attention, to the decrement of other *softer* and therefore less obvious factors (Clark and Moon 2001; Tolk 2003). In the military domain, technological interoperability has long been regarded not only as fundamental to interoperability between forces but also sufficient to enable close co-operation (Clark and Moon 2001). This has been reinforced by the network-enabled and network-centric doctrines on military activity, which have tended to emphasise the technological aspects of the network over the equally important social components (Tolk 2003). However, equipment is only one part of the interoperability spectrum, and merely networking organisations is as likely to lead to overload of personnel and worsen their performance as it is to lead to improvements (Hazel and Bopping 2006; Hutchins and Timmons 2007).

11.2.2 SOPs

The emergency services have SOPs for multi-agency interoperability via Airwave radios (ACPO 2010). The criteria for the invocation of interoperable voice communication include *'contributing to a common understanding and awareness of the situation'* (ACPO 2010, p. 4).

Airwave supports shared talkgroups by personnel from the three services at each level of command (bronze, silver, gold). The SOPs recommend that interoperability should not extend below the bronze level of command. This means that communications on the incident ground, between units from different agencies, is unlikely to be performed through Airwave talkgroups. In terms of *local* sensemaking, therefore, it is plausible to assume that the sharing of information and the co-ordination of activity are going to be performed through face-to-face and verbal communications, with radio communications being reserved for information sharing with each agencies' *bronze* command. Often, a multi-agency response will involve personnel on the incident ground forming cross-agency huddles (Figure 11.2). Such a huddle can share knowledge, set goals, etc., but only for those who participate in it. There is not an obvious means by which the information shared in a huddle can be distributed more widely to the incident. Further, a huddle only tends to be formed at a location where personnel from different agencies are operating in proximity and when the rhythm of the incident dictates that there is time to do so. In Chapter 10, we noted that information liaison officers carry information between

Figure 11.2 Cross-agency huddle at incident.

agencies, but their role can become challenged when the incident tempo or information load increases.

The interoperability SOPs are partly seen as necessary to *discourage improvisation* and to *maintain the integrity of command within each agency* (ACPO 2010). However, large-scale emergencies often demonstrate the requirement for innovation and improvisation, as we saw in Chapter 5. In response to the flooding around Walham electricity substation in Gloucestershire in 2007, the improvisation included the creation of new command roles (platinum command, pseudo silver, gold liaison), unorthodox scene management (fire and rescue service at Walham) and a complex C2 arrangement between Gloucestershire and Avon Fire and Rescue Services.

11.2.3 Training

'True interoperability is built on mutual understanding, familiarity and trust between the Emergency Services and partner agencies' (ACPO 2009, p. 27).

Association of Chief Police Officers (ACPO) (2009) hold that effective interoperability is achieved through training and exercising together; however, as Chapter 5 demonstrates, major incidents may involve working with entirely new organisations, as well as with unfamiliar personnel. Whilst multi-agency exercises may be useful in identifying operational issues and developing joint practices (ACPO 2010), it is unrealistic to expect local emergency services to conduct the regular large-scale training and exercises that would be required to generate familiarity in this manner – especially as it is difficult to devise exercises for what are unexpected and unprecedented events. Therefore, the question is how to foster trust within the *ad hoc* networks that are generated in response to emergencies, in order to enable them to effectively collaborate to make sense of the unique situation.

This also raises a question as to what is being trained in multi-agency exercises. Whilst it is essential to ensure that all technology systems are able to operate effectively together, and whilst it is important for all agents to understand their roles in the multi-agency SOPs, we feel that the issues surrounding sensemaking might receive less attention. We acknowledge that most exercises begin with the need to discover what is happening and to develop a response in light of this understanding, and that exercise scenarios also involve unexpected events to further challenge personnel. However, the manner in which agents understand (or make sense of) the scenario is not always tested, evaluated or measured as objectively as it could be.

11.2.4 Usage

ACPO (2009) holds that interoperability should be part of normal business, i.e. *'The response to emergencies should be grounded in the existing functions of organisations and familiar ways of working, albeit delivered at a greater tempo, on a larger scale and in more testing circumstances'* (ACPO 2009, p. 19).

This is seen as having already been established, as it will be conducted '…in line with existing local protocols between the Emergency Services' (ACPO 2009, p. 27). However, for organisations to closely collaborate in order to make sense of a major incident is likely to require more than merely switching to an interoperable radio talk group. As Chapter 5 of this book has demonstrated, during large-scale emergencies, organisations will continue to rely on their own C2 networks to make sense of and respond to the incident. It is unreasonable to expect personnel to shift to new ways of working just as the situation is at its most demanding. ACPO (2009) hold that each emergency service should retain their C2 structure, which is likely to mean that they will continue to gather, assess and act upon information separately.

11.3 Case Study: Initial Response to Terrorist Attacks on July 7, 2005

On the morning of July 7, 2005, four bombs were detonated on the public transport system in central London; three of the explosions took place in quick succession on London Underground trains, with the fourth detonating on a double-decker bus (Figure 11.3). About 52 commuters were killed, and over 700 were injured. The attack was designed to cause maximum disruption, and London's emergency services activated a major incident response. Despite their best efforts, the emergency services encountered problems in organizing their initial response in terms of a collaborative cross-agency effort. These difficulties were arguably attributable to problems of organizational control and information sharing (7 July Review Committee 2006).

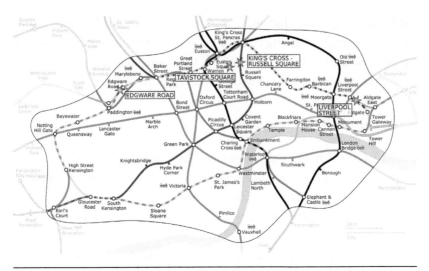

Figure 11.3 Map showing the locations of the July 7 explosions.

Whilst the service control rooms and resources at the scene were able to share information, and pre-existing response plans agreed by the services indicate that they should liaise with each other from the start of an incident (LESLP 2007), the emergency services faced problems in the co-ordination, the collection and the assessment of incident information. Response data presented in the Report of the 7 July Review Committee suggested that the reason for this may lie in the fact that the services were organising their responses using their individual command structures and information management systems, with little incident information being passed between them.

Figure 11.4 shows the elapsed minutes before each organisation involved in the response effort declared a major incident at the various scenes and emphasises the variation in response across organisations. A summary of the initial response to each incident is provided in Appendix A. The large disparity between the declaration times suggests that the services were unable to reach consensus on the nature of the incident during the initial response and were operating as three separate services. Additionally, as the number of major incidents increased, there is a decrease in response efficiency. From the incident summaries, it can be seen that on July 7, the efficiency of the emergency service response varied across the incident sites and over time; there were a number of deviations from an idealised response, including the following (Report of the 7 July Review Committee 2006):

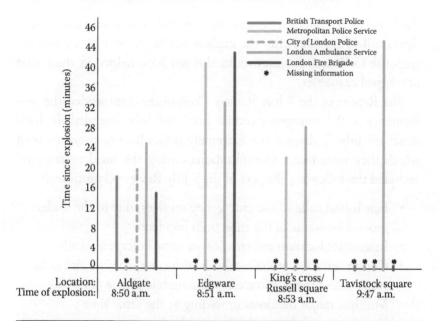

Figure 11.4 Time since explosion for emergency services to declare a major incident at the scenes, based on timings in the Report of the 7 July Review Committee.

- Delays in despatching resources
- Decreasing efficiency as the number of major incidents increased
- Failure to fully resource some sites
- Delays in declaring a major incident and establishing the associated command structure
- Equipment shortages
- Decisions taken outside of the command structure

In spite of this, the important issue is whether the emergency services were able to adequately achieve their primary goal during the immediate response, i.e. the evacuation and treatment of seriously injured casualties. The incident data indicate that the delays in implementing the incident responses ate into the *golden hour* for the seriously injured casualties who needed hospital treatment; for example, the first ambulances did not arrive at Edgware Road until 27 minutes after the explosion, and the site had not been declared a major incident by all three services until 43 minutes after the explosion (7 July Review Committee 2006). Similarly, at Russell Square, the first ambulance did not arrive until 37 minutes after the explosion, and the London Ambulance Service did not declare it to be a major incident until 45 minutes after the explosion. The absence of fire brigade response to Russell Square would also not have helped in the rescue of trapped casualties.

The Report of the 7 July Review Committee commended the performance of the emergency service personnel who attended the incidents on July 7, despite the extremely difficult circumstances with which they were faced. Complications during the incident response included the following (Report of the 7 July Review Committee):

- Few initial calls to the emergency services, due to the underground locations of the tube train incidents
- Inaccurate location information in some emergency calls
- Communication problems (both with emergency service radio systems and the overloaded public mobile phone network)
- Multiple major incidents unfolding at the same time
- Restricted and dangerous incident locations

11.4 Case Study: Initial Response to Boston Marathon Bombings (2013)

On April 15, 2013, two homemade bombs exploded near the finish line of the Boston Marathon (Figure 11.5), killing 3 people (including 8-year-old Martin Richard) and injuring some 264 people (16 of which required amputation of limbs). As the bombings affected an organised event, there was a well-resourced medical area (Station Alpha) at the finish line, with 16 ambulances and a large number of medical personnel on hand to treat marathon runners as they completed the course. The presence of so many medical personnel near the site of the explosion, together with fire and police personnel at the scene, meant that the initial triaging and treatment of the injured could be performed instantaneously.

A law enforcement co-ordination centre (analogous to silver command, Chapter 5) was established quickly to co-ordinate activity and prioritise response. Some 40 minutes later, a unified command centre (UCC; analogous to gold command, Chapter 5) was established to co-ordinate response. It should be noted that, whereas the gold command in the London bombings was primarily focussed on responding to the explosions and rescuing victims, the Boston Marathon UCC

THE CHALLENGES OF INTEROPERABILITY 177

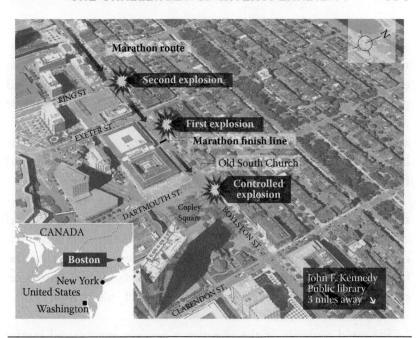

Figure 11.5 Location of explosions during Boston Marathon. Available at http://www.independent.co.uk/news/world/americas/map-how-the-boston-marathon-bombings-happened-8575248.html.

was focussing on both the medical response and the law-enforcement response to identify and apprehend the perpetrators. Whereas the medical response was clear and obvious to the public, some of the law-enforcement activity was somewhat confusing. For example, cell phones were taken from casualties who were admitted to hospital, in order to conduct a forensic examination of the phones and any images that they contained. For some of the casualties and their relatives, the taking of cell phones, belongings, clothing, etc., and the questioning in terms of the law-enforcement activity added to the distress of the bombing. Whilst it was clear to the police and law-enforcement personnel that the incident required a twofold response, the public might have been more concerned with their immediate medical needs. The law-enforcement perspective was further illustrated when another explosion occurred some seven minutes later (at 2:56 p.m.) in the John F. Kennedy Library. Whilst this later turned out to be an electrical transformer fire, it was initially linked to the two explosions at the marathon as a co-ordinated set of terror attacks. One can see how parallels could be drawn between the multiple explosions in London and the set of explosions in Boston.

By 2:57 p.m., the hospitals in the region had activated emergency operations centres (some had already activated such centres due to the marathon being run), following the notification (via radio and email) of a mass casualty incident (which they received around 2:53 p.m.). In order to manage the transfer of casualties from the scene to the hospitals, the UCC set up a second staging post and put out a request for private ambulance companies. Some 73 private ambulances responded. The second staging post had to be moved several times as reports and investigations of further threat were assessed.

An interesting coda to the dual response to the incident mentioned above was that the air monitoring reports (for radiological and chemical residue) were completed by 3 p.m. and shared with incident command on the scene, but this information was not passed to the hospitals. Although this might not have had a significant impact on response, it does illustrate how the collection of information can be stovepiped and shared on a *need-to-know* basis.

The rapid response of the emergency services was praised by the subsequent enquiry (7 July Review Committee 2006), which also noted that a 2012 tabletop training exercise, of the agencies that would be involved in managing the marathon, involved a scenario in which there was a mass casualty incident on the finish line.

11.5 Conclusions

All incidents are unique. We have presented discussions of bombings in London and Boston to illustrate the differences in how sensemaking could take place in response to major incidents. There are as many differences as there are parallels between these two incidents, and so a direct comparison would be foolish. However, we feel that there are two striking points to consider here. The first is that the nature of the incident, i.e. the semantic sense, that responders were addressing was immediately clear and shared by in Boston. (Even if the cause was not immediately apparent, it was clear that a specific event, an explosion, had occurred in a specific location and that this explosion had specific effects.) This meant that declaring a mass casualty (major) incident could be agreed unanimously and quickly. Part of the challenge in the London bombings lays in determining the location of the incident and the nature of the event. Making sense is about deciding what to respond to and where it is happening. The second is the deployment of resources. For the Boston bombings, emergency responders were on the scene and able to offer an immediate response. (And the recruitment of additional resources was achieved rapidly and efficiently.) For the London bombings, the deployment of resources was dependent on whether the incident was major or not and on getting vehicles through busy London streets to different locations. This question of when an incident is declared and who is involved in this declaration is considered in the next chapter.

12
SENSEMAKING AND ORGANISATIONAL STRUCTURE IN EMERGENCY RESPONSE

In major incidents, multiple agencies need to work together. This raises the question of interoperability (explored in Chapter 11) and highlights the challenge of ensuring that information sharing is well managed. In this chapter, a social network analysis of communications during the initial response to the July 7, 2005 bombings highlights the problems of adapting conventional emergency response practice to the demands of a major incident.

12.1 Introduction

The UK Emergency Services (ACPO 2009, 2010) views interoperability in terms of voice communications between equivalent layers of command, in response to exceptional circumstances (i.e. major incidents) that do not normally impact upon existing separate C2 structures. As such, the communications merely replaces the multi-agency *huddles* that are seen during the consolidation phases of major incidents with a shared talk group. Thus, the initial response phase of incidents will still be co-ordinated entirely separately through the existing C2 systems of each service. As we noted in Chapter 11, this view of interoperability is drawn from the US Department of Homeland Security model and is, we believe, fairly representative of the view of communications in emergency response internationally. From a sensemaking perspective, this approach to interoperability makes a number of assumptions regarding the nature of major incidents:

- That it will be obvious when an incident is a major incident
- That it will be obvious what the nature of the incident is
- That it will be obvious what information needs to be exchanged
- That the exchange of non-voice data between services is not critical to effective incident co-ordination

As we have noted in Section 1.1.1, these assumptions may often be the exception rather than the norm, and this means that a critical initial phase of any incident response lies in the challenge of determining what is happening. We see this challenge as the primary domain in which sensemaking occurs (although if the situation should radically shift, then a new phase of sensemaking could take place).

12.2 Sensemaking as a Social Process

The situation awareness of personnel from different services and agencies attending the same incident may be very different, due to different training, doctrine, tactical goals, priorities and role specialisation, such that the fundamental nature of the situation may be viewed differently. Additionally, the separate C2 networks (both organisationally and technically) sometimes lead to different information being passed along.

Comparing the two case studies in Chapter 11, the bombings in Boston were witnessed by personnel at the scene, and the initial response could be rapidly mobilised using resources that were already present. In London, the nature of the incident took time to establish, and the challenge of deploying resources was compounded by the confusion as to whether this was a single or multiple major incident and the precise location of each explosion. Furthermore, the alignment of the emergency response in Boston occurred immediately, whereas it took longer for this to happen in the London bombings (again, due to the nature of the incidents and the responses being managed by the individual agencies). It is worth noting that the Boston bombings did involve two parallel responses (one medical and one related to law enforcement) and that there is evidence of some slippage between these responses, e.g. in terms of how and where information was communicated. In the Walham flood case study (Chapter 9), the three emergency services operated different organisational structures, managing different geographic areas (e.g. counties versus regions), which lead to different priorities. They also had very different doctrine, training, procedures and powers and operated under separate budgets, administered through separate government departments and are even taxed differently. Personnel have different working conditions, and the organisations work to entirely different performance criteria, so whereas their overarching aim is the same (to protect life), their motivators during any given situation may cause them to prefer different responses. At lower levels of the organisation, common overarching goals can translate into different priorities and tactics; the Walham electricity substation discussed in Chapter 5 demonstrated how this can lead to tensions between the responding agencies. Even where there is a broad strategic consensus, common high-level goals and priorities do not necessarily translate into co-ordinated tactics and operations. As such, the establishment of clear, shared goals is a fundamental requirement for effective international co-operation.

The recognition of command itself may not be guaranteed. Where an organisation is perceived as having taken control of an incident away from other agencies, there is the risk of causing resentment, which may hamper the response effort and future working relationships. Given the importance of having all relevant information available from partner agencies when making decisions, the generation of resentment or competition amongst a coalition may be extremely damaging to the effectiveness of the joint response.

The discussion of the defence of the Walham electricity substation in Chapter 9 highlighted the fact that the unique features of the situation and environmental constraints required close co-operation between the responding agencies to make sense of the situation. The major hurdle in enabling a co-ordinated response came in terms of achieving a commonality over how the problem was to be viewed and the roles of each agency within the response. There were then difficulties in terms of communicating the requirements to achieve the solution, which were compounded by technological constraints and unfamiliarity between different agencies that rarely work together. The nature of the problem was thought to be self-evident, yet in reality, the various agencies viewed this *simple* problem very differently, and an intensive discussion was required to generate a shared understanding. From this discussion, it is interesting to consider how the manner in which communication takes place influences interoperability and, ultimately, how this may affect sensemaking.

The Report of the 7 July Review Committee concludes that better communication between the emergency services might have led to a faster implementation of a co-ordinated response to the emergencies. The report highlights the role that is played by technological factors in creating problems for the emergency services on the July 7, 2005 (7/7) bombings. For example, difficulties were experienced with the analogue emergency service radio networks, and reliance on the public mobile phone network by responders led to problems when it became heavily oversubscribed following the explosions.

High levels of demand on mobile phone network led to communication problems for the City of London Police, which implemented access overload control (ACCOLC) on the cell-phone network that is operated by the service provider O2, for 1 kilometre around Aldgate. The issue of whether to use ACCOLC to restrict access to the mobile phone network provides further indication that the emergency services were operating independently of each other on 7/7. The City of London Police appear to have taken the decision to implement ACCOLC at Aldgate without consulting with the other emergency services, despite the potential negative impact on their ability to communicate with resources at the scene. Also, the City of London Police reported that they were not aware of the earlier decision of the gold co-ordinating group not to use ACCOLC, indicating a lack of communication between the levels of command (7 July Review Committee 2006).

12.3 Analysing Network Structures and Interoperability

Figure 12.1 shows the social network diagram that is derived from the detailed descriptions in the Report of the 7 July Review Committee of all incidents on 7/7. The nodes represent agents who were involved in the response, such as individuals, control rooms and responding units; the links represent communications between the agents, with the width of the arrow and the associated numerical value indicating the number of communications. It is important to note that this diagram does not show all communications that took place during the initial response phase; there would have been

SENSEMAKING IN EMERGENCY RESPONSE

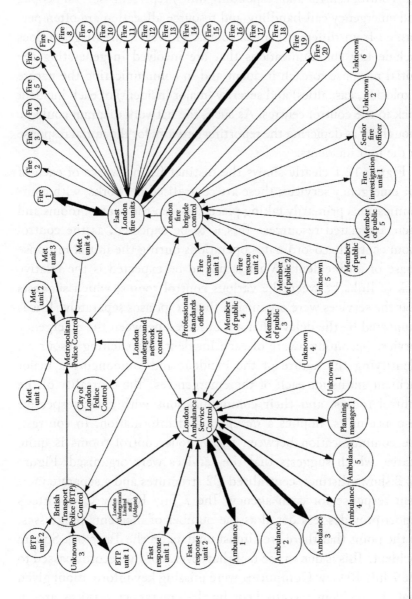

Figure 12.1 Social network for the initial response to all incidents on 7/7.

numerous communications between the various control rooms and responding units and between resources at each scene. The social network diagram also oversimplifies the C2 structure as each agent (i.e. control centres and responding units) represents several people, and emergency call-handling and resource allocation are often performed from different control rooms. However, the diagram does include key communications that are involved in generating the initial response to each incident and in communicating the nature (explosion, casualties) and severity (major incident) of each incident back to the control centres. As such, the network diagram could be thought of as depicting the reporting network for the initial response to the four incidents.

Figure 12.1 clearly shows the distinct C2 networks of each of the emergency services; these are centralised networks, with communications primarily taking place between the control rooms and their associated resources. This is to be expected, as the control room centrally co-ordinates operations during the initial response phase of an incident. What is not to be expected is the relative lack of links between the various control rooms, which suggests that the services were organising their responses separately. This is supported by the delays that are observed between the emergency services becoming aware of each incident, despatching resources, identifying the nature of the incident and announcing a major incident status for each of the emergencies. The centrality of the control rooms, and their communications with their respective response units, implies a *stovepiping* of information. In contrast, the communication between the various control rooms is quite sparse, which suggests that the services were organised. Figure 12.2 shows distinct, centralised C2 structures and a separate incident response for each agency. The 7 July Review Committee's report provides less detail as the number of incidents progresses, to the point that there are almost no data on the Tavistock Square incident. This is due to the fact that several of the records passed to the 7 July Review Committee were missing key information; given that the incident records kept by the emergency services are an official log of the response to an incident, this would indicate that the emergency services were not able to keep pace with the flow of information relating to the incidents.

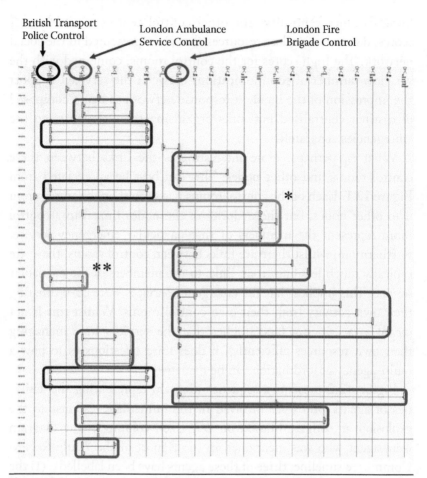

Figure 12.2 Sequence diagram showing the activity of different agencies on 7/7.

Considering the network in terms of centrality allows us to explore the relative prominence of each agent in the network and the standing of each node within a network. The methods for calculating centrality are described further by Houghton et al. (2006). A full set of analyses is presented in Appendix B. In order to determine the significant agents in this network, we rank agents by centrality scores and then define thresholds, in terms of the mean score, to define inclusion. From this analysis, the agents who exert most influence in the network and who have the shortest communication paths are the London Fire Brigade (Control), the British Transport Police (Control) and the Metropolitan Police Service (Control). The London Ambulance Service (Control) has high degree centrality but lower distance centrality. The other control centres received lower scores, due to having less resources that are represented in the social network. The high scores for the control rooms, combined with the very low number of communications between the services, support the impression of the incident response network as being composed of distinct centralised networks organising their responses to the emergencies separately.

What is, perhaps, more interesting is the links between these control rooms and other nodes. This can best be appreciated from Figure 12.1. Each of the control rooms has several links coming from it to other nodes, but the majority of the nodes linked to a control room represent the members of that particular emergency service. Furthermore, whilst there are links between control rooms, these are typically much less busy than links to the members of the particular service. This implies that the emergency services were co-ordinating their own activity from their own control rooms. Whilst this is not meant as a criticism (and implies that the services were managing their own resources efficiently), it does raise questions as to how a multi-agency co-ordination can be managed.

Figure 12.2 presents a sequence diagram representing the activities of agents who were involved in the immediate response to the explosion at Aldgate Station on 7/7. Each column in the diagram is an agent who was mentioned in the Report of the 7 July Review Committee timeline; three of these agents have been labelled – (1) the British Transport Police Control, (2) the London Ambulance Service Control and (3) the London Fire Brigade Control. The sequence diagram (Figure 12.2) represents activities over time, with each row in the diagram indicating a point in the incident timeline. Actions by individuals are represented as yellow boxes, and the horizontal lines joining them together indicate the communications between the agents. What the sequence diagram shows is that, whilst there were numerous communications during the initial response to the explosion, they were almost all within service and from the control room to the responding units. For example, the London Fire Brigade communications are all between the control room and the fire brigade resources that have been despatched to the scene. The same is true of London Ambulance Service and British Transport Police communications. Two exceptions to this are highlighted in orange: the first (marked *) shows the communications that were sent at 8:59 a.m. to all emergency services to attend the three underground incidents; the second (marked **) was a communication from the British Transport Police control to the London Ambulance Service control to request their attendance at the incident.

12.4 Conclusions

The 7/7 report concluded that more effective communications between the emergency services may have been beneficial in reducing the period of uncertainty surrounding each incident, resulting in a faster, more co-ordinated response (7 July Review Committee 2006). In terms of the social network issues that we have identified with the incident response, a common information environment may well have had a positive impact, shortening the delays between the various services becoming aware of each incident, despatching resources and identifying and declaring a major incident status. Once one service had identified the nature, location or scale of an incident, this information would become available to the other services, allowing them to immediately despatch resources. A common operating picture (COP) would also represent the current status of all incidents, allowing supervisors and higher command levels to identify problems as they developed. For example, the incidents at Russell Square and Tavistock Square had inadvertently been given the same rendezvous point for the resources that are sent to the scene; this meant that ambulances that had been despatched to Russell Square were redirected to Tavistock Square on arrival (7 July Review Committee 2006).

Observations with other emergency services provide an illustration of the most likely method of sharing information on 7/7: during multi-agency incidents in the West Midlands region, personnel from the different emergency service control centres must telephone each other and exchange incident reference numbers before passing on information verbally, as the incident management software systems for ambulance, fire and police are not currently compatible.

It is plausible to assume that the separate C2 networks operated by the emergency services are optimised for *routine* incidents, which often require a single service response (for example, a burglary or an individual medical emergency); the emergency services will co-operate on these small-scale incidents where necessary; however, they tend to be more straightforward, requiring less intensive collaboration (other than operational-level details). Major incidents tend to be large scale and complex, with high levels of uncertainty, due to their unexpected and sometimes unprecedented nature (Faggiano et al. 2011). Consequently, they require collaborative problem-solving, which in turn requires close co-ordination between the emergency services at all command levels. Our analysis of the 7/7 incident response data suggests that the separate command structures did not encourage the co-ordination of activity. Changes to the organisation of emergency service C2 and the way that incident information is captured, stored and shared that would create a common information environment may make the organisation of multi-agency responses easier.

The sharing of data could allow the creation of a COP across the emergency services and other local and regional agencies. This is discussed further in Chapter 13. During multi-agency operations, the use of a COP may be beneficial in speeding the development of an appropriate response; by combining the incident information that is received by each agency within a COP, the emergency services would be able to rapidly alert each other, resolve inconsistencies and compile a more accurate representation of the situation. Major incidents can progress quickly, with rapid changes that affect the level of risk to service personnel; with a COP, as soon as one service becomes aware of the scale and severity of the situation, this information could be made available to all services, who would adjust their response accordingly.

12.5 The Challenge of Sharing Information

In their discussion of the response of agencies to an earthquake in California, Harrald and Jefferson (2007) note that the US National Homeland Security Response system is predicated on the assumption that all parties develop and share situation awareness. They identify a number of problems that bedevilled response efforts (and that we feel are common across many emergency response scenarios).

12.5.1 Data and Information Semantics

During initial response, data and information are available from a broad range of sources, ranging from telephone calls from members of the public to news reports to messages from personnel on the ground. Crucial problems can arise when people employ similar terms in different ways. For example, if a bridge is described as *damaged*, does this mean that it is still passable or that it has collapsed? In the case studies that are considered in this book, the question of semantics can be studied at both the individual level, with people applying different meanings to the words that they use to describe risk or other salient concepts, and at an organisational level, with different organisations interpreting concepts in different ways, for example, <cordon>?

One approach to dealing with variation in semantics would be to ensure that all communication occurs face to face to allow clarification and consensus to be reached before further decisions are made. This is one of the benefits of the briefing huddle (see Chapter 10). If it is not possible for people to meet face to face, then another approach might be to perform some form of information triage prior to circulating it. This would involve a central communications post that collates information and then distributes messages in a standard format. Another approach might be to encourage the use of common vocabulary (as we discussed in Chapter 13).

12.5.2 Data Quality, Quantity and Timeliness

A further problem arises in terms of the processes for gathering data. Are the sources reliable (in terms of their veracity, or in terms of their timeliness)? Data could be true at some point in the past but no longer correct, e.g. a report of flood levels at 3 metres could have been correct at 2 p.m. but the water levels might have changed at 6 p.m. Even if the data can be trusted, there is a further problem in terms of knowing how much data are available (in terms of the range of sources that are available) and, more importantly, how much data the decision making needs to make a decision. Whilst this latter point might have the feeling of the proverbial length of a piece of string, in high-stress, highly dynamic situations, it is possible for decision makers to wait for more and more information; this form of *decision inertia* (Alison et al. 2015) can delay the decision making or potentially lead to erroneous decisions being made.

12.5.3 Data and Information Relevance

Data and information could be of high quality, consistent and timely, but there remains the challenge of determining whether they are relevant to a specific operation. In the next chapter, the ways in which relevant data and information can be defined and shared are explored.

13
Common Operating Pictures

One approach to supporting collaborative sensemaking is through the provision of information displays that can provide a shared view on the situation. This chapter considers the design, development and uses of common operating pictures. We illustrate how an incident ontology can be used as the basis for the design of such representations, and point out the distinction between a common operating picture (which could be used to provide an overview of all the information pertaining to an incident) and a common relevant operating picture (which presents a subset of the total information that is relevant to a given agency, role or activity).

13.1 Introduction

As we noted in Chapter 12, a significant challenge for multi-agency response arises from the need to share information among agencies. Whilst the challenge of sharing information is multi-dimensional, one commonly mooted solution is to use a common operating picture (COP) that is shared across agencies (Keuhlen et al. 2002; McNeese et al. 2006). The concept of a COP is similar to the *blackboard architecture* (Erman et al. 1980) in artificial intelligence; team members are able to post and retrieve information form a common repository to which they have equal access (Baber et al. 2013). In this way, everyone involved in an incident could view an up-to-date version of key information.

The emergency services recognise the importance of a COP and that a compatible technology can assist with its development:

> *A significant outcome from effective interoperability is the Common Operating Picture (COP). The COP is a single, identical understanding of current and relevant information which is shared between the Emergency Services and partner agencies.* (ACPO 2009, p. 14)

The NPIA *Guidance on Multi-Agency Interoperability* discusses the benefits of having a COP across responding agencies, but then goes on to highlight the importance of protecting sensitive and restricted information. It suggests that steps are taken, such as restricting access to communications networks and having sessions of certain meetings closed to non-security cleared personnel, though it acknowledges that this will affect the decision-making capacity of partner organisations (ACPO 2009). If organisations are not able to appraise themselves of all the relevant information prior to decision-making, then the combined response will not be fully co-ordinated, which would seem to defeat the aims of the COP.

13.2 Informing versus Understanding

Chapter 5 discussed how agencies with different backgrounds, training and experience could come to different conclusions on the appropriate response to an incident, based on broadly the same set of information. This has serious implications for multi-agency operations. If the intention is to use the COP to focus on the communication of critical incident information (i.e. location, type, size, casualties, etc.), then the various organisations *may* have a shared awareness of the characteristics of the incident but will still maintain their own assessments of the best response. This could result in a *macro-organisation* (i.e. one in which all agencies are joined) that is only superficially connected, where the intent is not shared and where it is not possible to interpret the actions of the other agencies.

Whilst it may be possible to represent information in a format that is understandable to different agencies, this may still allow for the development of different (and potentially contrasting) assessments of the situation and best method of resolution. Shared information displays may merely change the question from one of *what is that organisation doing?* to one of *why on earth are they doing that?*, unless some method of representing and communicating each agency's assessment of the situation and the most appropriate response can be found – something that is currently achieved through face-to-face discussions. As Wu et al. (2013) note, the multi-agency emergency response is particularly challenging because the people who are working together often do not fully appreciate the activity and roles of other agencies within the emergency plan being implemented. To address this issue, they propose that some means of representing the plan, agencies' roles and activities in a manner that can support collaboration between people in different locations. Such a representation can be considered a form of a COP. The use of a shared representation ought to enhance collaborative sensemaking, although, as we shall argue in this chapter, this depends on what is being represented, how it is being shared and how both the representation and the sharing can be incorporated into the other activities of the sense makers.

Only by sharing perspectives and intentions prior to initiating the response would a conflicting activity *on the ground* be avoided. It is not clear that the current approaches to the development of COPs provide the freedom for agencies to make such statements of intent within the COP, and thus they will not be captured. Furthermore, even if such technology were available, then there is the potential problem that incident management could become overtaken by *COP driving*, i.e. in which there is a need for demand feeding of the COP at the expense of efforts to deal with the incident itself. Obviously, this is an extreme position (but one that we have heard expressed in informal discussions with serving personnel) and highlights the need to develop clear approaches to the integration of the COP with command processes.

Figure 13.1 is taken from a system that is used by several UK police forces and shows the geographical information system (GIS) view, which can be used to present the locations and status of incidents and officers. However, this view is utilised relatively infrequently in the emergency dispatch of police officers, as the task is primarily one of matching the available resources to incidents.

Many examples of COPs appear to make the assumptions that (a) it is immediately clear what type of incident it is, (b) there is an obvious *ideal* response and (c) all relevant information is known. In part,

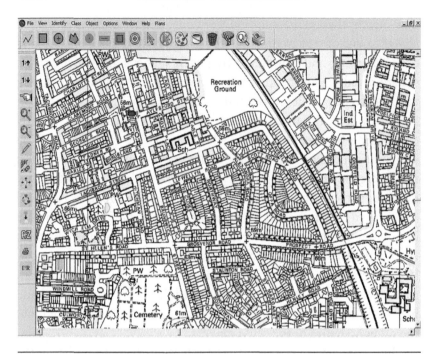

Figure 13.1 Emergency services map system.

this might simply be the result of using GIS to represent movement in space. One can easily indicate the location of an entity or group, and plot their movements. Further, one can provide some graphical means by which the level of confidence in the location, say, can be represented, e.g. colour or shape coding (although even this could prove problematic). This could mean that information that is not easily assimilated onto the map either needs to be represented somewhere away from the primary source or held in the minds of the planners. In either case, a close-knit, well-drilled team can easily handle such ambiguity, but the expansion of the team might lead to either problems of understanding or a significant burden to communicate and explain assumptions. It might also mean that the *shared map* begins to limit discussion purely to spatial rather than operational issues. However, spatial representations can only be expected to cover some aspects of operational command. This is one reason why there is much interest in the use of common data and knowledge structures. In the next section, we consider the use of (software) ontologies to provide such structures.

13.3 Ontologies for COP

As noted in Section 13.2, it might be beneficial to represent the information that needs to be collected during an emergency response and to use this representation to help manage both the incident response and the sharing of such information. Broadly, this is the aim of approaches to developing technology for incident response that make use of ontologies. In some studies, the sources of information extend beyond the emergency services and include crowdsourced information. The challenge is to define a means of structuring all of the information coming into the incident response system so that it can be easily processed and shared. The structured nature of an ontology could mean that problems of miscommunication and misunderstanding become reduced because all the information coming into the system has been triaged and processed into a common format. From this perspective, the COP becomes the means through which such a common format is displayed.

To illustrate this point, consider the acronym Casualties, Hazards, Access, Location, Emergency Services, Type (CHALET), which is used by some emergency service personnel to evaluate risk in incidents. Figure 13.2 takes this acronym and expands it in terms of an incident. In this case, the incident is an example that is taken from a training exercise that we observed in the Fire Service Training College, in which a road traffic collision between a car and a chemical tanker has led to the spillage of toxic chemicals. Each branch in this diagram represents a particular aspect of CHALET, and moving along each branch results in more details about a specific aspect of the incident. It might be assumed that all of this information is essential to the management of the incident. However, such richness of information might only be relevant to the silver commander who is managing the response. For each of the agencies working at the bronze level, it might be sufficient to have enough information to allow them to plan and co-ordinate that aspect of the response for which they are responsible. This introduces the distinction between a COP, which contains all the information about an incident, and a common relevant operating picture (CROP), which contains the information that is required for a specific agency, role or activity. In terms of the *push* of information, it might be useful to have a CROP for some of the responders.

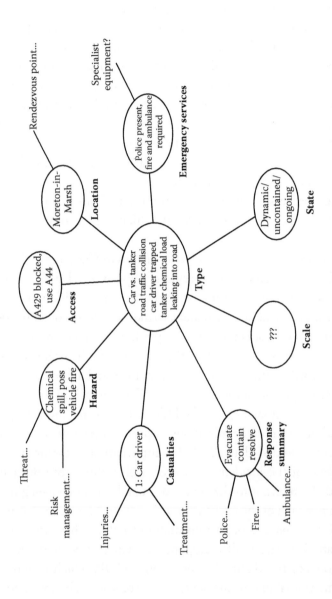

Figure 13.2 Mind-map representation of CHALET acronym.

COMMON OPERATING PICTURES 199

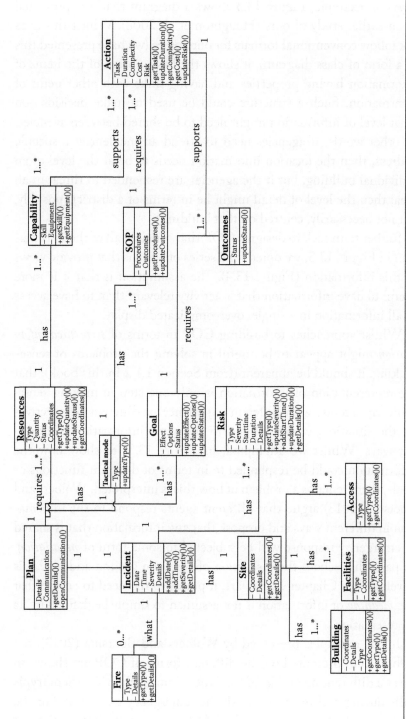

Figure 13.3 Class diagram of information in an incident.

As an example, Figure 13.3 shows a diagram that was presented in an earlier study of ours (Houghton et al. 2008). Whilst this does not follow conventional formats for ontologies (we have presented this as a form of class diagram), it shows the basic notion of the items of information having properties and having links with other items of information. Such a structure could be used to make decisions on what level of information might need to be shared between agencies. In other words, if agencies need to attend an incident at a specific address, then the location information needs to be at the level of an individual building, but if the agencies are responded to threat in an area, then the level of detail might be in terms of a district (possibly, but not necessarily, centred on that building).

Rather than seek to design a COP that contains all of the information in Figure 13.3, we opted for a series of screens that provide views of this information (Figure 13.4). The assumption is that it is more useful to have information that is activity relevant than to have access to all information in a single, overcomplicated display.

Whilst approaches to building COP in terms of *structured information* might appear to be useful in solving the problems of sense-making, it should be apparent (from Section 1.1.2 in this book), that the representation of information is only one step in making sense. Different agents, with different experience or different perspectives on the situation, could respond to the same information in different ways. Whilst we attempted, in Figure 13.3, to show how the information could be responded to in terms of different functions or goals, there remains a problem of how this is interpreted. Wolbers and Boersma (2013) argue that different agents respond to the information in different ways and propose that any information that is shared in emergency response will be subjected to some form of *negotiation*. In part, this echoes the discussion of common ground and how this develops (see Chapter 1). In part, it points to the need to ensure that the *meaning* of information is not assumed to simply be defined by its representation.

In the examples presented by Wolbers and Boersma (2013), the information presented on the different forms of COP are shown to carry a different meaning for the professionals who view them (typically during real responses or during emergency exercises). For the professionals, the content of the COP is not simply *information* but

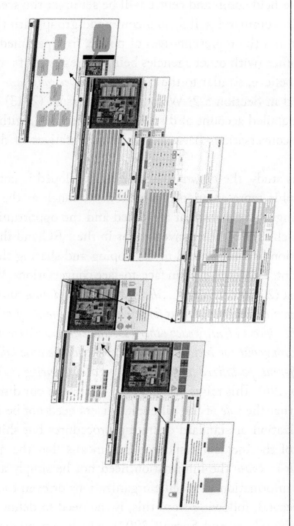

Figure 13.4 Examples of user interfaces designed to support Figure 13.3.

is also often interpreted as an invitation to act or, at the very least, to seek clarifying information from other sources. Their study relates to the Dutch emergency management system in which a call to the emergency response centre (ERC) can lead to the dispatch of agencies to the incident scene. If the incident is sufficiently challenging, a mobile field command centre will be set up at the scene, and the officer-in-command will form a command group with the field commander and the representatives of police, fire, paramedics and municipal office (with other agencies being represented if required). This is, we believe, similar to the concept of *silver command*, which is considered in Section 5.2. Wolbers and Boersma (2013) present a rich and detailed account of discussions in the ERC, with examples of the conversations between the representatives of different agencies.

From this study, the *picture* that is being developed is not necessarily a graphic representation of the scene so much as the overall understanding of the risks that are posed and the opportunities for action for each agency. The conversations in the ERC and the ensuing negotiation become crucial to developing and sharing the sense of the incident. However, even in face-to-face conversations, Wolbers and Boersma (2013) note that '...*there can be limited understanding of what consequences information has for the action and needs of other professionals. The officers identify each other's specialized knowledge and roles differently throughout the incident response operation because officers can represent different specialized clusters of organizations during the response operation*' (p. 196). This raises two issues of interest to our discussion. The first is that the *role* of the different officers need not be fixed to their organization or standard operating procedures but shifts over the course of the incident response. This means that the question of *who needs to know* the information need not be simply a matter of aligning information to specific organizations or even to specific roles. The second, following from this, is the need to define *actionable knowledge* (Cross and Sproull 2004), which can progress current tasks. To some extent, this reflects Landgren's (2004) distinction between *committed interpretation* and *committed action*, in terms of striking a balance between the information that people need to know in order to appreciate the scope of the situation and what they need to know in order to take action in response to that situation. In terms of COP, the question then arises as to how best to define information as having the potential to lead to actionable knowledge (particularly when the COP is not the result of a face-to-face discussion but is being enacted through technology).

13.4 COP as a Representation of the State of the World or as a Collaborative Planning Tool?

Within the emergency services, a number of systems that are analogous to COPs have been developed, to optimise the use of limited resources across the numerous new incidents that are continually unfolding and for this information to be shared across multiple control rooms and supervisors.

Figure 13.5 shows the incident list for a region of a UK police force, which displays the information that the dispatcher requires in order to manage responsive policing: (a) the location of the incident, (b) the type of incident, (c) the priority and (d) the current status. Another window behind this view shows a list of resources and their current status. By switching between the two views, the dispatcher is able to match the available resources to the highest-priority incidents and only refers to a GIS screen when a specific information on the location is required (for example, access points to a property).

A GIS-based COP could include information on the intended actions of partner agencies and the expected development of the situation over time (e.g. fire, riot); this changes the COP from merely presenting the current (or, more accurately, last known) state of the incident to one that projects the future state, enabling the emergency services to ensure that their actions align with each other over time. This then takes the COP away from being a visual representation of information and towards a tool for collaborative planning activity. For example, Wu et al. (2013) designed and evaluated a geo-visualization tool to support collaborative sensemaking. The user interface allowed users to search and annotate maps and then share these annotated maps. The evaluation involved a simple task in which a family needed to be rescued and then provided with shelter. The participants, working in groups, needed to determine which of four options was the best location for a shelter. Individual team members focussed on specific issues, such as mass care, public works and environmental conditions,

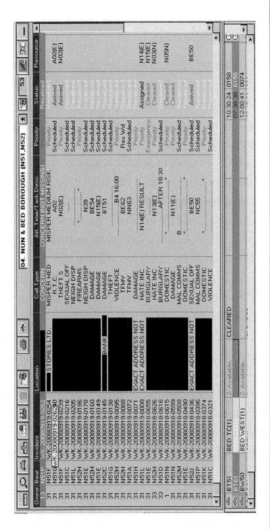

Figure 13.5 Incident log list.

and could then contribute to a shared map. The evaluation compared teams using the software with teams working with paper in a face-to-face manner. The results showed that teams using the software tended to *pull* more information from a shared space than those working face to face (although there was no difference in the *push* of information). Push involves placing information in the shared space in anticipation of other users needing it (which could be more efficient than having a two-step *request for information* followed by *providing information*). The observation that these push acts increased in frequency over trials (for both conditions) suggests that teams were better able to anticipate information needs.

In work that we have reported elsewhere, we have compared small teams on a simple collaborative task. In this task, people working on separate computers had to collaborate to build a jigsaw puzzle. The task was performed either with a central co-ordinator (whose role was to collect and distribute pieces to participants) or with open sharing of pieces. In essence, we were replicating a hierarchical and an edge network (discussed in Chapter 5). In this study, we demonstrated that there was little difference in performance for simple puzzles (i.e. with <10 pieces), but there were significant differences between the two conditions when the number of pieces increased (Baber et al. 2007? network-enabled capability through innovative systems engineering paper). Surprisingly, perhaps, the hierarchical teams performed better than the edge teams when the puzzles became more complicated. The difference in performance was simply down to the increasing number of communications for participants in edge teams (in terms of requesting and responding in sharing of pieces). In the hierarchical condition, the pieces were pushed by a central node. On the one hand, this provides further support for the idea of *pushing* information as an efficient means of managing activity. On the other hand, it highlights the challenge of managing through a centralised network. This could create information bottlenecks in which information overload could limit efficiency, or it could create stovepiping in which information becomes held by a single agency rather than shared. The withholding of information could arise from the knowledge and understanding of a given agency and, as we saw in the initial stages of response in the July 7, 2005 bombings, could become exacerbated when the situation becomes highly stressful. In such circumstances, the ability to select information for one's own activity could become paramount, and the selection of information that might be usefully pushed to other agencies becomes an additional task. One way in which this *additional task* could be made easier is to have a means by which information is tagged for its relevance to particular agencies, roles and activities.

13.5 Situation Space versus Decision Space

For the most part, COP is concerned with providing information that allows emergency personnel to gain an overview of the situation to which they are responding. As Drury et al. (2009) have pointed out, it is not only vital to appreciate the situation as it unfolds but also equally important for decision makers to be able to understand the consequences of actions that they might order. '*An optimal COA [course of action] is always extremely situation-dependent and is often sensitive to conditions beyond decision-makers' control*' (Drury et al. 2009, p. 537). In other words, there is likely to be a gap between the manner in which a situation is described (e.g. in terms of timeliness, quality of information) and the precision of information that is needed to make an optimal decision. On the one hand, this points to the need to *satisfice* in emergency response (which, in turn, calls to mind the discussions of individual decision making in Chapter 2). On the other hand, it highlights what Hall et al. (2007) term the *situation space–decision space* gap. Drury et al. (2009) demonstrate that providing decision makers with information relating to the decision space (i.e. visualising the cost of alternative courses of action) led to superior decision making (more correct, less ambiguous) and greater confidence in the decision makers (compared to not providing such information). The primary means by which cost is displayed to the decision maker is a box plot, showing the median, limits and interquartile range of the cost of options on some common scale (Pfaff et al. 2010). For example, cost might be defined in financial terms or in terms of the time to respond (or some composite function of terms).

13.6 COP as Product versus COP as Process

The discussion in Section 13.5 has presented COP as an artefact, or as a representation of information that can be updated and shared by different agencies. However, the notion of a COP as the product of sensemaking (either in terms of the focus of sensemaking activity or as the result of such activity) is not a complete description of its role in collaboration. First, parties might not pay full attention to the COP (perhaps due to the tempo of the incident to which they are responding, perhaps due to the information overload that they are experiencing or perhaps simply due to not understanding what is being presented on the COP) and verbally request information that they could retrieve from it (Paley et al. 2002). In this case, rather than the COP being the sole source of information, it is seen as one source amongst many. It is possible that some emergency responders would prefer to receive a verbal confirmation of the status of an incident than to consult the content of the COP; this might be particularly the case in a high-stakes response. This highlights the challenges of integrating the COP into the decision-making processes. Second, the COP will only be as good as the information that is displayed on it. This can result in the *product driving* the *process*, i.e. in terms of the need to maintain, update and author the content of the COP. This could easily become a specialised role in the incident response system. Not only can the COP drive the process, in terms of requiring constant information; it can also drive the process in terms of shaping the discussion of the incident. For instance, as we have noted, when a GIS is used to represent an incident, then the responder might describe the aspects of the incident spatially (and might be unable to include the aspects that are not easy to represent spatially). This is a variation on the common knowledge effect (Gigone and Hastie 1993), in which a group discussion is far more likely to focus on shared information that is common to all members of the group than to address information that might be known by a smaller number of group members.

Although, of course, this would depend on a host of factors pertaining to the situation, group dynamics, etc., there are several reasons why groups might focus on shared information (Kerr and Tindale 2004). This could range from the social validation, which arises from the discussion information that everyone else knows, has an opinion on and does not require explaining or justifying, to the likelihood that shared information is easier to bring to mind than novel or obscure information (which means that the discussion around such information is cognitively less demanding than the discussion on new information).

From this perspective, a potential danger in overreliance on using a COP to manage incidents could be that it would provide a distorting lens through which to view the incident: a lens that only shows information that can be represented in the format of the display and that shows only information that all users agree on. Whereas this might be suitable to incidents that are unfolding in a predictable manner, it could cause problems in incidents where it becomes vital to consider the *left-field* solution to the problem. In other words, it might not always be possible to exhaustively describe the ontology of a situation. This means that there might be *new* information that does not fit the ontology. The question then becomes whether one modifies the ontology or whether one represents the information *outside* of the ontology.

13.6.1 Routine Emergencies

During routine emergencies, the incident management system (IMS) performs a central role as a record of events, a resource for action and a means of communication between the call handlers and the controllers. It also provides a frame to structure sensemaking and represents the key incident features. At the same time, the digital radio network can facilitate collaborative sensemaking between response officers, by enabling mutual monitoring and providing an open forum for discussing incidents. As such, the IMS and the use of the digital radio as a forum for collaboration already come close to representing a COP for routine emergencies. However, there is currently a significant information bottleneck between the controllers and the responding officers (who do not have data access), as well as a limited opportunity for collaborative use of the radio, both of which restrict the controller's ability to allocate incidents intelligently and the ability of response officers to co-ordinate their actions. With minor adjustments to emergency service communications networks, it should be possible to better use the capabilities of the digital communications equipment to produce an environment that supports greater collaborative sensemaking during incident responses.

With the Airwave digital radio network, it is possible to have *nested* talkgroups (Heikkonen et al. 2004). One (or more) talkgroups could be used as dedicated forums for officers to collaborate during incident responses, whereas another talkgroup maintains the ability for control to share important information with everyone (Figure 13.6). This would free up the controller's talkgroup for urgent incident allocation, officer safety and officer inquiries. The responding officers would then be able to communicate in a less-formal open forum, enabling them to pool their knowledge and experience and to co-ordinate activity, without compromising critical broadcasts. Emergency broadcasts would automatically switch to the controller-monitored talkgroup, ensuring that they are picked up and responded to. Airwave also allows the transmission of data and therefore would support the use of data terminals in patrol vehicles. Figure 13.7 shows a data terminal inside a Helsinki Police patrol vehicle, which connects to a digital communications network operating on the same standard as the UK Airwave system. The

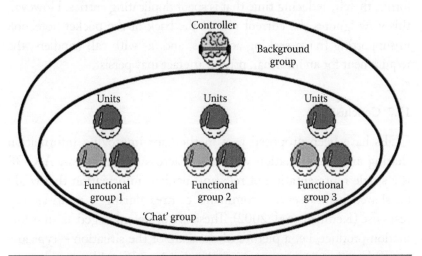

Figure 13.6 *Nesting* of talkgroups through use of scanning and talkgroup prioritisation.

Figure 13.7 Data terminal in a Helsinki Police patrol vehicle (2011).

data terminal in Figure 13.7 shows a GIS-based interface, which is used to mark the position and status of response vehicles and active incidents.

This information would be beneficial to responding officers trying to work out who is best positioned to respond to an incident, whilst access to the IMS and other police databases would allow them to conduct their inquiries directly, rather than having to go via the controller and thereby reducing radio traffic. At the same time, direct access to information systems could have negative impacts on incident response sensemaking. For example, this could reduce the ability for officers to monitor one another's inquiries and proactively engage in collaborative sensemaking. It may also be seen as removing the requirement for the pocket notebook, as information can now be entered directly into the IMS or crime forms, thereby reducing time that is spent duplicating entries. However, this view ignores the current role of the back of the pocket notebook in supporting frame-seeking activities, and, as with call handlers, the requirement for an informal, private artefact may persist.

13.7 Conclusions

COPs have been proposed as a method for improving information sharing and co-ordination within and across organisations. A COP is a single representation of relevant incident information that could be shared across service command centres during a multi-agency response (Keuhlen et al. 2002). These are often envisaged as an information product, i.e. a picture of the state of the situation – typically a GIS-based system for marking the positions of incidents and assets (McNeese et al. 2006). This book suggests that a lack of common ground is likely to preclude simply implementing a shared GIS, or an inter-organisational version of the IMS. Similarly, the notion of a COP that uses translation layers or filters to only show certain information to certain users – a *CROP* (Flentge et al. 2008) – is also flawed, as it fails to acknowledge the important role of collaboration in making sense of an incident, assuming that the nature of the problem and the associated solution are clear. An alternative view is that of the COP as an integral part of C2 processes that represents the combined incident knowledge space (i.e. a frame generation support tool) and enables the development of a shared understanding of an incident (McNeese et al. 2006).

14
Discussion

Sensemaking takes place when there is a gap between what is known and what needs to be done. It involves those processes in which information is collected, collated and sifted in order to define a situation with sufficient clarity, to allow a course of action to be selected. Very often, these processes involved interaction and collaboration not just with other people but also the artefacts that are provided an appropriated for work. Sensemaking is made difficult in part by the complex, dynamic, uncertain situations that people encounter, and in part by the contradictions, cross-purposes and constraints from the other people and the artefacts that are involved in those situations.

14.1 Introduction

This book presents a study of sensemaking in emergency response. Sense is made through an iterative process of collecting, collating and sifting information through verbal, physical and electronic representations of situation-relevant information. Thus, sensemaking goes beyond interpretation of a situation (Weick 1995; Maitlis and Christianson 2014). This means that, contrary to the assumptions, which seem to inform many of the artefacts that are developed to support or enhance sensemaking, it is not simply a matter of labelling the salient features of a situation and communicating or recording these features. If we are to maximise the benefits to be gained from technology, we need to design them not to impose their own form of making sense but rather to support the sensemaking activity of the people who will use them. As noted in Chapter 2, sensemaking is underpinned by the relationship between data and the frameworks that are used to conceptualise the problem. From this perspective, other people and the dialogues used to develop and manage common ground become critical to sensemaking. Equally important are the artefacts that people use to represent data and support collaboration between people.

The book offers a novel approach of sensemaking as distributed cognition. This approach comprises the three perspectives of (1) making sense with artefacts, (2) making sense through artefacts and (3) making sense through collaboration. Case studies from the emergency response command and control (C2) domain are presented, which describe sensemaking as distributed cognition during routine and major incident responses. The findings are summarised in Section 14.1.1, after which the remainder of this chapter discusses the implications of these findings in relation to sensemaking and distributed cognition theory, models of C2 and the emergency services domain.

14.1.1 *Sensemaking during Routine Emergencies*

Sensemaking during routine incident responses was found to be concerned with framing the problem; once an incident has been defined in terms of a recognisable *type*, SOPs can be applied in order to guide the process of resolving it. This implies that sensemaking ends when a course of action can be identified. In some instances, this could involve little or no explicit sensemaking; a routine and familiar event occurs, and a standard operating procedure (SOP) and can be selected and applied. In such circumstances, there is little need to engage in a deeper analysis of the situation. In other situations, the selection of the course of action could require a continuous revision. The concept of flexecution, considered briefly in Chapter 4, highlights how plans can become resources for replanning. This can occur when there is not a unique and unambiguous course of action to apply to the situation. This might occur in complex, dynamic domains where commanders continually need to re-evaluate the situation.

The process of framing the situation is a distributed cognitive activity involving multiple agents from across the C2 system, supported by several key artefacts. In Chapter 7, we noted how *informal artefacts* support rapid frame-seeking, questioning and elaboration, before formal artefacts (with a high distribution potential) are used to present the frame-seeking *product* and enable communication with other agents within the network.

Agents within the C2 network concentrate on specific elements of the sensemaking problem, with their output forming the basis of action for the next. Thus, the outputs become resources for action for future activity. In order to *bridge the gap* of inconsistent data and violated expectations, sensemaking becomes a collaborative activity; this is supported by technology such as the Airwave digital radio network, which allows for talkgroup monitoring and (occasionally) can become an open forum for collaboration and co-ordinated action.

The agents within the C2 system make use of rapid, highly compact, formalised communications in the entries in artefacts and during radio communications, which leaves much of the meaning and relevance of information as implicit. This is underpinned by an extensive common ground between agents, formed through shared training, experience and a common purpose. Even during collaborative reframing of more complex incidents, communication is still defined in terms of the established language and procedures, i.e. routine incident sensemaking is a culturally defined activity (Weick 1995). This indicates that sensemaking in routine incident response appears to function as a community of practice, which is enabled by the artefacts that are available to them.

14.1.2 Breakdown of Sensemaking during a Major Incident

As with routine emergencies, from the case study in Chapters 9 through 11, multi-agency major incident sensemaking is concerned with framing the problem, in order to identify the appropriate SOPs to apply in response. However, the *un-ness* (Hewitt 1983) of multi-agency major incidents results in more fundamental questions surrounding the nature of the problem and the solution, which transcend the agency-specific common ground that is associated with familiar routine emergencies and requires inter-agency collaboration. This could take the form of an exploration network; however, the responding agencies were initially found to maintain the individual organisational structures and social processes that are associated with communities of practice, drawing on standard frameworks and SOPs to make sense of events and consequently forming incomplete (and agency-specific) pictures of the incident.

The case study in Chapter 9 showed that such a *stovepiped* position was reinforced by the lack of artefacts, which, within the fire and rescue service, hampered the ability to delegate C2 tasks – adding significantly to the fire and rescue commander's workload. Although the bronze commander stated that the incident plan was all in his head, Chapter 5 argues that it is more probable that he made use of people and physical elements within the environment to represent information and act as resources for action. By delegating tasks and making regular rounds of the site, the bronze commander was able to draw on the various individuals to support sensemaking activity. However, an inability to track off-site issues that could not be represented by physical objects or people within the site meant that these tasks were harder to remember, with the result that some were not completed.

In terms of inter-service collaboration, the lack of shared artefacts to represent the *problem space* or compatible communications technology meant that physical proximity between commanders was required, limiting the ability for distributed agencies both to monitor one another's actions and to contribute to the sensemaking process. This would have exacerbated the difficulties associated with identifying important features of the incident that fell outside of an individual agency's incident response lexicon.

In the case study, liaison officers (LOs) were used to resolve interoperability shortfalls, both within and between C2 networks. The LOs provided an important role in translating one organisation's sensemaking output into terms that are meaningful for another, which would not have been replicated merely by providing access to shared artefacts. However, the use of LOs led to the fragmentation of the fire and rescue C2 network, leading to a loss of sensemaking continuity (mission command) across the various command levels. The role of LOs as inter agency interfaces also falls short of interagency collaborative sensemaking, which was required to form new interpretations of the incident.

14.2 Reflections on Sensemaking Theories

We have been developing the notion of sensemaking as distributed cognition. This approach argues that cognitive processes involved in framing problems are mediated through interactions with artefacts and other agents. Within this approach, three perspectives: making sense with artefacts, making sense through artefacts and collaborative sensemaking. These perspectives combine distributed cognition theories with current approaches to sensemaking, namely the process of representation construction (Pirolli and Card 2005), the mapping of data to frames (Klein et al. 2006a,b) and collaborative search-after-meaning (Weick 1995). Current sensemaking approaches focus on specific aspects of cognitive activity within defined contexts and are limited in their wider utility. By adopting a systems-level view, the new approach views these sensemaking approaches as interrelated processes and draws on their strengths to produce a comprehensive description: the data–frame model provides the generic process by which sensemaking takes place at all levels of an organisation; the representation construction approach views frames as external to the individual, which supports individual and collaborative sensemaking; collaboration provides a means for novel frame development. Rather than merely combining these theories, the new approach extends beyond them into a holistic view of sensemaking as a technologically mediated and socially distributed cognitive activity. As Section 6.1 describes, this approach was successfully applied to the study of incident response activity, with Chapters 4 and 5 providing detailed descriptions of how systems-level sensemaking activity is mediated through distributed cognition processes.

14.2.1 Individual Sensemaking

We propose that, in complex emergencies, there is no single *expert*; rather, sensemaking should be a social process that actively involves all relevant participants. During the defence of Walham electricity substation (Chapter 9) and during the initial response to the July 7, 2005 bombings, experts from different backgrounds drew different conclusions from the same information precisely because they were not engaging in a collaborative sensemaking process, as they lacked the social and technological means to do this easily. This case study suggests that the view of commanders as individual experts may be actively damaging, as they will recognise some familiar elements of an incident, without understanding the significance of unfamiliar but equally important factors, causing them to apply existing (inappropriate) frames. In the data–frame model's own terms, the distinction is between *frame-seeking* and *frame-defined* sensemaking activity. In the former, a problem space is defined and explored; in the latter, a situation is recognised, and an SOP can be applied. Whilst the latter might lead to initiating a faster response to the problem, in complex and situations, it might also lead to *recognising* the wrong solution. The case studies and vignettes presented depict many of Klein et al.'s (2007) types of sensemaking: (a) frame-seeking, (b) frame-defined data collection, (c) questioning the frame and (d) reframing, lending weight to their argument that sensemaking is a complex group of associated activities.

14.2.2 Artefact-Driven Sensemaking

Artefacts support sensemaking by people at all levels in emergency response, as they forage for information, develop a model of what was/had happened, elaborate this and then *presenting* a final product that is used by other agents in the system. This description broadly aligns with the *representation construction* sensemaking approach that is presented by Pirolli and Card (2005) and Attfield and Blandford (2011). It also supports the wider application of the representation construction view beyond the low-tempo activities that were investigated in previous studies. In terms of *how hypotheses/schemas/frames are generated*, pre-defined incident categories provide the framework for the incident, which are tested by the call handler and the response officers through interactions with artefacts and members of the public. The flexibility of informal, private artefacts allows them to play an important role in supporting frame-seeking activity prior to the use of formal, public artefacts for frame-defined data collection and sharing. Disparities were also uncovered between the intended and actual use of artefacts during sensemaking, often in instances where either the process or artefact design do not fully support the sensemaking activity.

The representation construction approach, however, can be criticised for failing to account for social processes. Whilst sensemaking can be an individual process, the activities described by Pirolli and Card (2005), and Attfield and Blandford (2011) are not entirely stand alone, in that their products (intelligence/investigation reports) are used as resources for subsequent actions by other agents (e.g. criminal proceedings). In the same way, the sensemaking activities of the call handler sit within the wider incident response process and the *product* (the IMS log) initiates the next phase of activity by the controller. The highly compact, formal language used in shared artefacts, such as the IMS log, is only possible because of the common ground that exists amongst the agents within the system. Similarly, the products of the legal investigations observed by Attfield and Blandford (2011) are likely to be incomprehensible to those outside the judicial system. This also suggests that artefacts normally used to support activity in one agency may not help inter-agency collaboration due to lack of common terminology or SOPs preventing the establishment of common ground. Thus, in terms of representation construction as a process of developing a sensemaking *product* for others to act on, social processes (Hutchins' [1995b] cultural heritage) are an important factor to recognise when investigating the use of artefacts to support sensemaking.

This book lends further support to the process that is described in Klein et al.'s (2006a,b) data–frame model. Responding to emergencies large and small is about making sense of complex and uncertain events. Responding appropriately requires an understanding of the nature of the problem, yet sensemaking is inextricably linked to action and interpretation, meaning that incident response activity is a constant process of the refinement of both the understanding of the incident and the associated response. The framework used to make sense of an incident defines subsequent action and interpretation. Thus, making sense of an incident becomes a question of framing the problem. However, the findings of this book do not support all of the assertions that are applied to the data–frame model, as is discussed.

14.2.3 Collaborative Sensemaking

This book demonstrates that responses to emergencies – both small and large – feature collaborative sensemaking processes. Examples of sensemaking processes from across the incident response C2 system display the characteristics that are identified by Weick (1995), which is described in Chapter 4. During a collaborative incident response activity, the crucial sensemaking element is the identification of, and agreement on, the most plausible understanding of the event (i.e. the frame). Once the frame has been established, it is then largely possible to apply SOPs to deal with the response. This underlines the point that sensemaking is not an end in itself but merely an input into the wider C2 processes that are involved in incident responses (i.e. decision making, planning).

Weick (1995) views sensemaking as firmly grounded within social activity, taking the organisation as the level of analysis. Whilst it is clear that collaboration is taking place during the responses to both routine emergencies and major incidents, it is not correct to view sensemaking as being fundamentally rooted in social activity. To do so would risk missing important features of sensemaking activity that take place at a lower level of granularity.

Weick's (1995) view of sensemaking as single organisations engaged in culturally defined disagreements may have more to do with the types of environment that are studied in the development of the approach, rather than the actual phenomena itself. This book lends weight to this argument, demonstrating that collaborative sensemaking takes place not only during *normal operations* (i.e. single agencies dealing with a defined problem space) but also during *exceptional situations* (i.e. multi-agency responses to complex, novel and uncertain events), which more closely represent Umapathy's (2010) view of sensemaking as involving coalitions of groups with different worldviews. The book indicates that the notions of communities of practice and exploration networks are relevant and useful descriptors of collaborative sensemaking, not least because they offer some explanation for how collaborative sensemaking may take place (which is largely missing from this approach). However, the major incident case study demonstrates that whilst a particular inter-agency collaborative style may be preferable for specific incident conditions, it cannot be taken as a given that organisations will reorganise themselves to meet this *ideal* arrangement.

The main problem with Weick's (1995) view of sensemaking as a systems-level activity is that it offers little explanation of how sensemaking takes place. This book has successfully applied Klein et al.'s (2006a,b) data–frame model to the problem, arguing that artefacts are able to act as shared representations (of the frame), thereby facilitating collaboration, including amongst geographically and temporarily distributed agents. Chapter 4 provides an account of the use of artefacts to store and represent information, to support reflection and reinterpretation and to cue activity and the communication of findings,

14.2.4 Sensemaking as Distributed Cognition

The approach to sensemaking as distributed cognition presented in this book is not intended to represent a fully comprehensive sensemaking model that applies to all situations; however, it does provide a coherent holistic approach to describing sensemaking within complex C2 systems. Klein et al. (2006b) and Weick (1995) begin their descriptions of sensemaking *in* the situation, describing the sensemaking processes of individuals and organisations during and after the event. Similarly, the descriptions of representation construction provided by Pirolli and Card (2005) and Attfield and Blandford (2011) begin after the *problem* has been detected. This book goes some way towards addressing the question of what happens before this, by taking the opportunity to study incident response C2 from across the incident life cycle, demonstrating that collaborative sensemaking begins long before a commander arrives at the scene, as well as playing an important role in incidents without a *commander*. This suggests that, to an extent, the C2 system is less *bridging a gap* and more *starting from scratch* with every new incident. This book demonstrates that incident response sensemaking is influenced by a range of factors, including the situation, organisational structures and procedures, common ground (or lack of) and supporting artefacts (or lack of).

In discussions of crime scene examination, Baber (2013) notes that '…*it is not always apparent where the act of "cognition" is situated*' (p. 131). Similarly, this study of incident response C2 raises the question of where within this process sensemaking takes place. Formal organisational structures and doctrine would suggest that there are central sensemaking foci for both routine (controller) and major incident (commander) responses. Previous sensemaking theories would suggest that it is either (a) *within-the-head* of key individuals (Klein et al. 2006b), (b) arises during collaboration (Weick 1995) or (c) a process of artefact transformation (Pirolli and Card 2005; Attfield and Blandford 2011). The three perspectives of making sense with artefacts, making sense through artefacts and collaborative sensemaking applied in this book indicate that sensemaking during incident response activity is a complex, interwoven combination of individual, collaborative and artefact-based activity, with specific elements coming to the fore at various stages of the process and depending on the situation.

Prior studies have tended to focus on moments of crisis that bring sensemaking to the fore, but this does not mean that sensemaking does not take place during slower tempo activities, i.e. where a process is not only geographically and socially distributed, but also temporally extended. For example, one of the authors (RM) witnessed something akin to this process taking place within the incident response domain. At the start of a shift, response officers are briefed by their sergeant; this briefing will include an informal team debrief of the incidents of their last shift – where officers will relay the narratives of particularly interesting incidents to one another and update them regarding individuals of note. This activity is combined with a slower-tempo formal sensemaking loop, which comprises intelligence messages that are given during the brief, that are collated and *sanitised* (the origins of the information are concealed) by analysts from various sources, including formal incident records within the IMS and crime files. Frequently, these formal and informal organisational sensemaking loops are combined during briefings, whereby officers pool their various insights on individuals and events that are named in formal briefings, to collaboratively form their own interpretations of what the criminal fraternity might be up to. The approach taken within this book offers a means for investigating less obvious but nevertheless important systems-level sensemaking activities.

As an explanation of sensemaking as a systems-level activity, the *sensemaking as distributed cognition* view proposed in this book is vulnerable to the argument that it has only been applied to incident response C2. This, in turn, suggests further study to evaluate the relevance of this approach to sensemaking during planned C2 activity (e.g. air traffic control), as well as to other domains where *control* is not only latent but also where it is entirely absent, such as the artefact-based mediated activity of civilian members of social networking sites as they form *ad hoc* networks to collaboratively make sense of and respond to large-scale crises (cf. Duffy and Baber 2013).

Finally, given the unique nature of major incidents and the absence of normally available command support in the case study presented in Chapter 5, further study of sensemaking during major incidents is warranted, not least to begin to redress the imbalance that is represented by the volume of commander-centric studies.

14.3 Revisiting C2

Whilst this research has focussed specifically on *reactive* incident response (which is only one element of the purpose of C2), it does enable reflection on current thinking on the nature of C2. Returning to the generic process model of C2, the notion of *command* and *control* as activities, rather than roles, is supported by this book. For example, during routine emergencies, control functions are performed by a number of agents, and whilst there is no active commander input,* a large part of the command activity is performed by response officers. The notion that units on the ground are in control of the incident is at variance to traditional notions of C2 and is surprising, given that emergency service organisational structures appear to show highly centralised and rigid networks. Given the opportunity, response officers will collaborate to make sense of incidents, forming *ad hoc* sub groups of the overall network. Thus, in some ways, the police incident response C2 network described in Chapter 5 offers hints of how it could function as an edge organisation, if access to relevant information and the ability to form *ad hoc* collaborative networks were not constrained. In military terms, the activity looks like a *mission command*, in which a broadly defined intent cascades from the commander to local units, who will flesh this out into courses of action that are appropriate to the local situation. The challenge is to maintain high-level accountability whilst also allowing local adaptability. This points to an ongoing challenge for sensemaking researchers, which is how the *sense* of a situation varies not only across agencies (with their differences in SOP, training and knowledge) but also hierarchically within an organisation (in terms of different measures of performance). Whilst the initiation of a gold command (in UK parlance) can bring some of these issues to the fore, the manner in which satisfying different *senses* of

---

* Commanders are present within the C2 system but rarely take an active part in routine emergency responses.

the situation can create conflict and dilemmas. This was illustrated, in Chapter 9, by the discussions surrounding the delivery of diesel for the generators at Walham.

The edge organisation (Alberts and Hayes 2003) does not give any suggestion as to how or why agents within a large, heterogeneous network would form *ad hoc* collaborative groups. This book demonstrates that rapid *ad hoc* collaboration requires a pre-existing community of practice, whereas during large, complex, multi-agency operations, disparate organisations with little common ground do not have an obvious incentive or understanding of the requirement to engage collaboratively.

When compared to the case studies presented in this book, the generic C2 model (Chapter 5) fails to give any sense of the uncertainties that are involved in incident resolution or the effort that is required to frame the problem, having only the two discrete steps of *determine mission* and *determine events*. In reality, during incident responses, the problem is rarely neatly defined from the start, and this book suggests firstly that a number of ongoing activities are hidden behind these two steps (potentially including the development and maintenance of the problem frame, mission command, inter-agency collaborative networks and building common ground), but also that determine mission and determine events are interwoven activities, as the developing problem frame shapes further sensemaking. This model also takes a linear view of the process of responding to an incident, whereas the case studies presented in this book demonstrate that sensemaking is a fundamental C2 activity that is conducted both throughout the life of an incident and throughout the C2 system. These criticisms may stem from the fact that the model is predominantly based on studies of normal operations and activity-planning cycles, rather than reactive operations or crisis response.

14.4 Future Directions for Incident Response

Routine emergencies and major incidents share a number of common features; however, this research has identified important differences in the nature of the problems that are faced and the associated sensemaking demands. Consequently, different COP solutions would be required to improve sensemaking during the different types of incident – suggesting further avenues for research that could offer valuable academic and pragmatic insights.

14.4.1 *Major Incidents*

Major incidents display many of the features that have been linked to exploration networks (Burnett et al. 2004) in that a temporary arrangement of individuals from diverse backgrounds with little shared experience are required to collaborate intensively, in response to a complex, unique and poorly defined crisis. However, major incident C2 structures are built on single-service command hierarchies, and once established, the multi-agency response is not geared towards the context-rich collaboration that is required to explore the problem and develop novel frameworks, but concentrates instead on applying frameworks based on prior experience. As a result, agencies may only engage in collaboration once a crisis point is reached, resulting in the C2 system reacting to events, rather than proactively working to understand and anticipate the causal elements and potential outcomes. This may help to explain the causes of some of the problematical themes of multi-agency crisis response identified in Chapter 1 relating to co-ordination, information sharing, mutual awareness and integration. Major incidents can require the emergency services to closely co-operate with a wide range of other agencies, public bodies, private companies and the military, so these are issues that they will continue to face. During multi-agency operations, the emergency services therefore require the organisations, social processes and supporting technologies to enable the rapid collaborative development of an agreed frame for the problem and the associated response.

The COP suggested for routine incident responses relies on the single-service community of practice, and so would not be suitable for use by *ad hoc* multi-agency networks that may have very little common ground at the start of an incident. Instead, what is required is a means for distributed agents to be able to articulate and represent their perspectives on the key features of the incident in order to engage in a collaborative process of novel frame generation. The preference for informal, unstructured artefacts to support frame-seeking described in Chapter 4 would seem to suggest providing some form of digital equivalent that allows multiple users (both co-located and distributed) to add, manipulate and question information in a highly informal manner. For example, a range of digital artefacts exist that mimic the behaviour of paper and whiteboards. These combine the advantages of informal, unstructured analogue artefacts with the distribution potential of digital artefacts. Combined with verbal communications (for example, through a radio talkgroup), this would enable users to identify connections between information fragments and establish points of commonality and thereby begin to develop some form of concept map for an incident. Once a frame had been defined, the various agencies would then be able to apply the most appropriate SOPs to resolving the problem whilst maintaining the option to revisit and reinterpret the incident frame should the need arise.

Whilst interactions between the command layers of a service are currently supported by the use of digital radio communications and IMSs, it is argued that they would also benefit from the use of a shared, informal concept-mapping tool. The role of gold, silver and bronze command during major incidents is – at least at the start of the incident – largely concerned with framing the problem and ensuring continuity of purpose across the command levels. Such a tool would also help to ensure that the mission command is maintained during the life of what may be very large and complex incidents – particularly where the situation requires an adaptation of the C2 network (such as making use of aid from neighbouring services). A simple digital artefact could therefore be used to rapidly develop and share the incident frame between command levels and across agencies. In a similar manner to the mutual monitoring that is seen in Chapter 4, this may have the added benefit of cueing individuals to *push* information to one another, as the visibility of the problems being worked on across the distributed network would suggest *who needs to know what*.

Developing a COP that can support multi-agency collaborative sensemaking as an exploration network does not automatically mean that agents will behave accordingly, especially given that the participants will be starting off from within their service-specific communities of practice and may not see the requirement to discuss an apparently obvious incident. Collaboration is effortful, being outside of the normal response process, and organisations may be tempted to concentrate on their own activities when their workload dramatically increases during times of crisis. Therefore, there is also a requirement to develop *exploration network thinking*, i.e. to encourage command staff to actively engage with other agencies and to articulate (avoiding technical jargon), question and explore the implicit assumptions that underpin service-specific perspectives. Given that major incidents can involve a broad array of public and private organisations, many of which will have had no prior training or involvement in incident response activity, the *onus* will be on the emergency services to lead this collaborative effort, which should begin by making explicit the roles, capabilities, constraints and expectations of all of the organisations that are involved.

14.4.2 Emergency Services Interoperability

The findings from this research raise doubts over the current UK Emergency Service programme of interoperability improvement (cf. Chapter 1). This initiative risks constraining collaborative sensemaking, as it includes restrictions on wider incident monitoring and discourages improvisation (NPIA 2010). The programme also fails to address data interoperability issues, and so information sharing between the services will remain effortful, rather than becoming a part of normal practice. This initiative appears to be founded upon a flawed major incident doctrine (LESLP 2007; NPIA 2009) that underestimates the level of interaction and collaboration that is required to make sense of even relatively simple major incidents. This book suggests that the intended course of action for interoperability is likely to perpetuate the problems that are associated with incident responses. Both the routine and the major incident case studies discussed in this book demonstrate that intensive collaboration is often required in order to make sense of and resolve highly uncertain situations, and that different forms of technological support will be required, depending on the nature of the problem that is faced.

14.5 Final Words

Sensemaking is needed when there are gaps in the understanding of individuals confronting a novel situation. At an individual level, bridging these gaps can involve recollecting prior experiences and knowledge to determine an appropriate response to the situation. Such activity can be supplemented by the use of artefacts that are available to the person. However, such artefacts can also serve to constrain or confuse the activity. At an organisational level, there is a need to develop common ground in terms of the nature of the situation and, equally, in terms of the nature of the gaps, prior to developing consensus for (or, at least an appreciation of) a response to the situation. We argue that the perspective offered by distributed cognition provides a beneficial and useful means of reviewing and considering the manner in which individuals, groups and the artefacts available to them engage in sensemaking in response to incidents and emergencies.

References

ACPO. 2009. *Guidance on Emergency Procedures*. London: National Police Improvement Agency.
ACPO. 2010. *Standard Operating Procedure Guide on Police to Police and Inter-Agency Airwave Interoperability*. London: National Police Improvement Agency.
Adams, F. and Aizawa, K. 2008. *The Bounds of Cognition*. Oxford, UK: Wiley-Blackwell.
Alberts, D. S. and Hayes, R. E. 2003. *Power to the Edge: Command and Control in the Information Age*. Washington, DC: DoD Command and Control Research Program.
Alison, L., Power, N., von den Heuvel C., Humann, M., Palasinki, M. and Vrego, J. 2015. Decision inertia: Deciding between least worst outcomes in emergency responses to disasters, *Journal of Occupational and Organizational Psychology*, 88, 295–321.
Artman, H. and Garbis, C. 1998. Situation awareness as distributed cognition. In T. R. G. Green, L. Bannon, C. P. Warren and J. Buckley (Eds.), *ECCE 9: Proceedings of the Ninth European Conference on Cognitive Ergonomics*, Le Chesnay, France: European Association of Cognitive Ergonomics (EACE), 151–156.
Ascoli, G. A., Botvinick, M. M., Heuer, R. J. and Bhattacharya, R. 2014. Neurocognitive models of sensemaking. *Biologically Inspired Cognitive Architectures*, 8, 82–89.
Attfield, S. and Blandford, A. 2011. Making sense of digital footprints in team-based legal investigations: The acquisition of focus. *Human–Computer Interaction*, 26 (1–2), 38–71.
Auf der Heide, E. 1989. *Disaster Response: Principles of Preparation and Coordination*. St Louis, MO: Mosby-Year Book.

Baber, C. 2013. Distributed cognition at the crime scene. In S. J. Cowley and F. Vallée-Tourangeau (Eds.), *Cognition Beyond the Brain: Computation, Interactivity and Human Artifice*, London: Springer, 131–146.

Baber, C., Smith, P., Cross, J., Hunter, J. E. and McMaster, R. 2006. Crime scene investigation as distributed cognition. *Pragmatics and Cognition, 14* (2), 357–385.

Baber, C., Stanton, N. A., Houghton, R. J. and Cassia, M. 2008. Hierarchical command, communities of practice, networks of exploration: Using simple models to explore NEC command structures. Realising Network Enabled Capability (RNEC NECTISE), University of Leeds, UK.

Baber, C., Stanton, N. A., Atkinson, J., Mcmaster, R. and Houghton, R. J. 2013. Using social network analysis and agent-based modelling to explore information flow using common operational pictures for maritime search and rescue operations. *Ergonomics, 56*, 889–905.

Baron, R. M. and Misovich, S. J. 1999. On the relationship between social and cognitive modes of organization. In S. Chaiken and Y. Trope (Eds.), *Dual-Process Theories in Social Psychology*, New York: Guilford, 586–605.

Bartlett, F. C. 1932. *Remembering: A Study in Experimental and Social Psychology*. Cambridge, UK: Cambridge University Press.

Becerra-Fernandez, I., Xia, W., Gudi, A. and Rocha, J. 2008. Task characteristics, knowledge sharing and integration, and emergency management performance: Research agenda and challenges. *Proceedings of the 5th International Conference on Information Systems for Crisis Response and Management (ISCRAM2008)*, 5–7 May, Washington, DC, United States, 88–92.

Bennett, J., Bertrand, W., Harkin, C., Samarasinghe, S. and Wickramatillake, H. 2006. *Coordination of International Humanitarian Assistance in Tsunami-Affected Countries*. London: Tsunami Evaluation Coalition.

Bjørkeng, K. 2010. Sensemaking in practice: Reducing and amplifying equivocality through storytelling. *Organization Learning, Knowledge and Capabilities Conference 2010*, 3–6 June, Boston, 1–23.

Blandford, A. and Wong, B. L. W. 2004. Situation awareness in emergency medical dispatch. *International Journal of Human–Computer Studies, 61*, 421–452.

Boin, A. 2004. Lessons from crisis research. *International Studies Review, 6* (1), 164–194.

Boin, A. and T' Hart, P. 2003. Public leadership in times of crisis: Mission impossible? *Public Administration Review, 63* (5), 544–553.

Boin, A. and T' Hart, P. 2007. The crisis approach. In H. Rodríguez, E. L. Quarantelli and R. R. Dynes (Eds.), *Handbook of Disaster Research*, New York: Springer, 42–54.

Bolstad, C. A. and Endsley, M. R. 2003. Measuring shared and team situation awareness in the army's future objective force. *Proceedings of the Human Factors and Ergonomics Society 47th Annual Meeting*, Santa Monica, CA: HFES, 369–373.

Bourbousson, J., Poizat, G., Saury, J. and Seve, C. 2011. Description of dynamic shared knowledge: An exploratory study during a competitive team sports interaction. *Ergonomics, 54* (2), 120–138.

Boyd, J. R. 1996. The essence of winning and losing. *Unpublished lecture notes.* Available at https://fasttransients.files.wordpress.com/2010/03/essence_of_winning_losing.pdf.

Brown, A. D., Stacey, P. and Nandhakumar, J. 2008. Making sense of sensemaking narratives. *Human Relations, 61,* 1035–1062.

Burnett, M., Wooding, P. and Prekop, P. 2004. Sense making—Underpinning concepts and relation to military decision-making. *9th International Command and Control Research and Technology Symposium,* Copenhagen, Denmark, 14–16 September, Washington, DC: CCRP, 2–20.

Cabinet Office Civil Contingencies Secretariat. 2003. *Dealing with Disaster,* 3rd edition. London: Her Majesty's Stationery Office.

Carroll, J. M., Kellogg, W. A. and Rosson, M. B. 1991. The task–artifact cycle. In Carroll, John M. (Ed.), *Designing Interaction: Psychology at the Human–Computer Interface,* Cambridge, UK: Cambridge University Press, pp. 74–102.

Chase, W. G. and Simon, H. A. 1973. Perception in chess. *Cognitive Psychology, 4,* 55–81.

Chi, M. T. H., Feltovich, P. J. and Glaser, R. 1981. Categorization and representation of physics problems by experts and novices. *Cognitive Science, 5,* 121–152.

Chua, A. Y. K., Kaynak, S. and Foo, S. S. B. 2007. An analysis of the delayed response to Hurricane Katrina through the lens of knowledge management. *Journal of the American Society for Information Science and Technology, 58* (3), 391–403.

Clark, H. H. 1992. *Arenas of Language Use.* Chicago, IL: University of Chicago Press.

Clark, A. 2001. *Mindware: An Introduction to the Philosophy of Cognitive Science.* Oxford, UK: Oxford University Press.

Clark, A. and Chalmers, D. J. 1998. The extended mind. *Analysis, 58,* 10–23.

Clark, T. and Moon, T. 2001. Interoperability for joint and coalition operations. *Australian Defence Force Journal, 151.*

Clark, H. H., Schreuder, R. and Buttrick, S. 1983. Common ground at the understanding of demonstrative reference. *Journal of Verbal Learning and Verbal Behavior, 22* (2), 245–258.

Cohen, M. S. 1993. The naturalistic basis of decision biases. In G. A. Klein, J. Orasanu, R. Calderwood and C. E. Zsambok (Eds.), *Decision Making in Action: Models and Methods,* Norwood, NJ: Ablex, 63–99.

Cornelissen, J. P. 2013. Sensemaking under pressure: The influence of professional roles and social accountability on the creation of sense. *Organization Science, 23,* 118–137.

Crichton, M. T., Flin, R. and Rattray, W. A. 2000. Training decision makers–Tactical decision games. *Journal of Contingencies and Crisis Management, 8,* 208–217.

Cross, R. and Sproull, L. 2004. More than an answer: Information relationships for actionable knowledge. *Organization Science, 15,* 446–462.

de Jaegher, H. and di Paolo, E. 2007. Participatory sense-making: An enactive approach to social cognition. *Phenomenology and the Cognitive Sciences, 6,* 485–507.

de Marchi, B. 1995. Uncertainty in environmental emergencies: A diagnostic tool. *Journal of Contingencies and Crisis Management*, 3, 103–112.

Dervin, B. 2003. From the mind's eye of the user: The sensemaking of qualitative–quantitative metholodology. In B. Dervin, L. Foreman-Wenet and E. Lauterbach, *Sense-Making Methodology Reader: Selected Writings of Brenda Dervin*, Cresskill, NJ: Hampton, 269–292.

Donnellon, A., Gray, B. and Bougon, M. G. 1986. Communication, meaning, and organized action. *Administrative Science Quarterly*, 43–55.

Drury, J. L., Klein, G., Pfaff, M. S. and More, L. D. 2009. Dynamic decision support for emergency. *IEEE HST'09: Technologies for Homeland Security*, New York: IEEE, 537–544.

Dubrovsky, V. J., Kiesler, S. and Sethna, B. N. 1991. The equalization phenomenon: Status effects in computer-mediated and face-to-face decision-making groups. *Human-Computer Interaction*, 6, 119–146.

Duffy, T. and Baber, C. 2013. Measuring collaborative sensemaking. *Proceedings of the 10th International Conference on Information Systems for Crisis Response and Management*, 12–15 May, Baden-Baden, Germany, 561–565.

Dynes, R. R. 1970. *Organized Behaviour in Disaster*. Lexington, MA: Heath Lexington Books.

Eccles, D. W. and Tenenbaum, G. 2004. Why an expert team is more than a team of experts: A social-cognitive conceptualization of team coordination and communication in sport. *Journal of Sport and Exercise Psychology*, 26, 542–560.

Endsley, M. 1995. Toward a theory of situation awareness in dynamic systems. *Human Factors*, 37, 32–64.

Endsley, M. and Jones, W. M. 1997. *Situation Awareness Information Dominance & Information Warfare*. Dayton, OH: Logicon Technical Services Inc.

Endsley, M. R. and Robertson, M. M. 2000. Situation awareness in aircraft maintenance teams. *International Journal of Industrial Ergonomics*, 26, 301–325.

Environment Agency. 2007. Case study: 2007 summer floods. Available at http://www.environmentagency.gov.uk/commondata/acrobat/infrastructurestudy_1917458.pdf (accessed March 2008).

Erman, L. D., Hayes-Roth, F., Lesser, V. R. and Reddy, D. R. 1980. The Hearsay-II speech-understanding system: Integrating knowledge to resolve uncertainty. *ACM Computing Surveys (CSUR)*, 12, 213–253.

Evans, P. and Wurster, T. S. 2000. *Blown to Bits: How the New Economics of Information Transforms Strategy*. Boston: Harvard Business School Press.

Faggiano, V., McNall, J. and Gillespie, T. T. 2011. *Critical Incident Management: A Complete Response Guide*. Boca Raton, FL: CRC Press.

Faisal, S., Attfield, S. and Blandford, A. 2009. A classification of sensemaking representations. *CHI '09 Extended Abstracts on Human Factors in Computing Systems*, New York: ACM, 4751–4754.

Fields, B., Wright, P. and Harrison, M. 1996. Designing human system interaction using the resource model. In L. Yong, L. Herman, Y. Leung and J. Moyes (Eds.), *Proceedings of the First Asia-Pacific Conference on Human–Computer Interaction*, 25–28 June, Singapore: Information Technology Institute, 181–191.

Flentge, F., Weber, S. G., Behring, A. and Ziegert, T. 2008. Designing context-aware HCI for collaborative emergency management. *CHI Workshop on HCI for Emergencies*, New York: ACM.

Flin, R., Pender, Z., Wujec, L., Grant, V. and Stewart, E. 2007. Police officers' assessment of operational situations. *Policing: An International Journal of Police Strategies and Management*, 30, 310–323.

Flor, N. V. and Hutchins, E. L. 1991. Analyzing distributed cognition in software teams: A case study of team programming during perfective software maintenance. In J. Koenemann-Belliveau, T. G. Moher and S. P. Robertson (Eds.), *Empirical Studies of Programmers: Fourth Workshop*, Norwood, NJ: Ablex Publishing Corporation, 36–64.

Fraher, A. L. 2011. *Thinking through Crisis: Improving Teamwork and Leadership in High-Risk Fields*. Cambridge, UK: Cambridge University Press.

Gigerenzer, G. 1996. On narrow norms and vague heuristics: A reply to Kahneman and Tversky (1996). *Psychological Review*, 103, 592–596.

Gigone, D. and Hastie, R. 1993. The common knowledge effect: Information sharing and group judgment. *Journal of Personality and Social Psychology*, 65, 959.

Gioia, D. A., Thomas, J. B., Clark, S. M. and Chittipeddi, K. 1994. Symbolism and strategic change in academia: The dynamics of sensemaking and influence. *Organization Science*, 5, 363–383.

Gloucestershire Constabulary. 2007. *Gloucestershire Water Emergency 2007 Chief Constable's Memorandum to Environment, Food and Rural Affairs Committee*. Retrieved November 2012 from: http://www.gloucestershire.police.uk/forcepublications/Downloads/item8613.pdf.

Gorman, J. C., Cooke, N. J. and Winner, J. L. 2006. Measuring team situation awareness in decentralized command and control environments. *Ergonomics*, 49, 1312–1325.

Greater London Authority. 2006. *Report of the 7 July Review Committee*. London: Greater London Authority.

Hall, D. L., Hellar, B. and McNeese, M. 2007. Rethinking the Data Overload Problem: Closing the Gap between Situation Assessment and Decision Making, Proc. of the 2007 National Symposium on Sensor and Data Fusion (NSSDF) Military Sensing Symposia (MSS), McLean, VA.

Harrald, J. and Jefferson, T. 2007. Shared situational awareness in emergency management mitigation and response. *40th Annual Hawaii International Conference on System Sciences*, New York: IEEE, 23–23.

Hazel, G. and Bopping, D. 2006. Linking NCW and coalition interoperability: Understanding the role of context, identity and expectations. In *Human Factors in Network-Centric Warfare Symposium*.

Heath, C. and Luff, P. 2000. Team work: Collaboration and control in London Underground line control rooms. In C. Heath and P. Luff (Eds.), *Technology in Action*, Cambridge, UK: Cambridge University Press, 88–124.

Heikkonen, K., Pesonen, T. and Saaristo, T. 2004. *You and Your TETRA Radio*. Helsinki, Finland: Edita Prima Inc.

Hemmingsen, B. H. 2013. The quick and the dead: On temporality and human agency. In S. J. Cowley and F. Vallee-Tourangeau (Eds.), *Cognition Beyond the Brain*, London: Springer, 93–112.

Hewitt, K. 1983. *Interpretations of Calamity from the Viewpoint of Human Ecology*. Boston: Allen and Unwin.

Heylighen, F., Heath, M., and Van, F. 2004. The Emergence of Distributed Cognition: A conceptual framework. In *Proceedings of Collective Intentionality IV*. Siena, Italy.

Hoffman, R. R., Shadbolt, N. R., Burton, A. M. and Klein, G. 1995. Eliciting knowledge from experts: A methodological review. *Organizational Behavior and Human Decision Processes*, 62, 129–185.

Hollan, J., Hutchins, E. and Kirsh, D. 2000. Distributed cognition: Toward a new foundation for human–computer interaction research. *ACM Transactions on Human–Computer Interaction*, 7, 174–196.

Houghton, R. J., Baber, C., Mcmaster, R., Stanton, N. A., Salmon, P., Stewart, R. and Walker, G. 2006. Command and control in emergency services operations: A social network analysis. *Ergonomics*, 49, 1204–1225.

Houghton, R., Baber, C. and Chaudemanche, E. 2008. Integrating human factors with systems engineering: Using WESTT to design a novel user interface for incident command systems. In *Proceedings of the Human Factors and Ergonomics Society 52nd Annual Meeting*, Santa Monica, CA: HFES, 1944–1948.

HM Government. 2005a. *Emergency Preparedness: Guidance on part 1 of the Civil Contingencies Act 2004, its associated Regulations and non-statutory arrangements*. London: Cabinet Office.

HM Inspectorate of Constabulary. 1999. *Keeping the Peace: Policing Disorder*. London: HMIC. Availalble at http://www.homeoffice.gov.uk/docs/pol dis.html.

Hutchins, E. 1995a. How a cockpit remembers its speeds. *Cognitive Science*, 19, 265–288.

Hutchins, E. 1995b. *Cognition in the Wild*, Cambridge, MA: MIT Press.

Hutchins, S. G. and Timmons, R. P. 2007. Radio interoperability: There is more to it than hardware. Monterey, CA: Naval Postgraduate School, Department of Information Sciences.

Jenkins, D. P., Salmon, P. M., Stanton, N. A., Walker, G. H. and Rafferty, L. 2011. What could they have been thinking? How sociotechnical system design influences cognition: A case study of the Stockwell shooting. *Ergonomics*, 54, 103–119.

Jentsch, F. G., Salas, E., Sellin-Wolters, S. and Bowers, C. A. 1995. Crew coordination behaviors as predictors of problem detection and decision making times. *Proceedings of the Human Factors and Ergonomics Society Annual Meeting*. Santa Monica, CA: Human Factors and Ergonomics Society, 1350–1353.

Johannesen, L. 2008. Maintaining common ground: An analysis of cooperative communication in the operating room. In C. P. Nemeth (Ed.), *Improving Healthcare Team Communication: Building on Lessons from Aviation and Aerospace*, Aldershot, UK: Ashgate, 179–203.

Johnston, R. 2005. *Analytic Culture in the US Intelligence Community: An Ethnographic Study*. Washington, DC: Central Intelligence Agency, Center for Study of Intelligence.

Kapucu, N. 2008. Collaborative emergency management: Better community organising, better public preparedness and response. *Disasters, 32*, 239–262.

Kendra, J. M. and Wachtendorf, T. 2003. Elements of resilience after the world trade center disaster: Reconstituting New York City's Emergency Operations Centre. *Disasters, 27*, 37–53.

Kerr, N. L. and Tindale, R. S. 2004. Group performance and decision making. *Annual Review Psychology, 55*, 623–655.

Keuhlen, D. T., Bryant, O. L. and Young, K. K. 2002. *The Common Operational Picture in Joint Vision 2020: A Less Layered Cake*. Norfolk, VA: National Defense University Joint Forces Staff College.

Khalilbeigi, M., Bradler, D., Schewizer, I., Probst, F. and Steimle, J. 2010. Towards computer support of paper workflows in emergency management. *Proceedings of the 7th International Conference on Information Systems for Crisis Response and Management (ISCRAM2010)*, 2–5 May, Seattle, WA, 1–7.

Kirsh, D. 2013. Thinking with external representations. In S. J. Cowley and F. Vallée-Tourangeau (Eds.), *Cognition Beyond the Brain: Computation, Interactivity and Human Artifice*, London: Springer, 171–194.

Kirsh, D. and Maglio, P. 1994. On distinguishing epistemic from pragmatic action. *Cognitive Science, 18*, 513–549.

Klein, G. 2006. The strengths and limitations of teams for detecting problems. *Cognition, Technology and Work, 8*, 227–236.

Klein, G. 2007. Flexecution as a paradigm for replanning, part 1. *IEEE Intelligent Systems, 22*, 79–83.

Klein, G. 2011. *Streetlights and Shadows*. Cambridge, MA: MIT Press.

Klein, G. 2013. *Seeing What Others Don't: The Remarkable Ways We Gain Insight*. London: Nicholas Brealey Publishing.

Klein, G. and Crandall, B. 1996. *Recognition-Primed Decision Strategies*. ARI research note 96-36, US Army Research Institute for the Behavioural and Social Sciences, Alexandria, VA, Fairborn, OH: Klein Associates.

Klein, G., Ross, K. G. and Moon, B. M. 2003. Macrocognition. *IEEE Intelligent Systems, 18*, 81–85.

Klein, G., Pliske, R. and Crandall, B. 2005. Problem detection. *Cognition, Technology and Work, 7*, 14–28.

Klein G., Moon, B. and Hoffman, R. R. 2006a. Making sense of sensemaking II: A macrocognitive model. *IEEE Intelligent Systems, 21*, 88–92.

Klein G., Moon, B. and Hoffman, R. R. 2006b. Making sense of sensemaking I: Alternative perspectives. *IEEE Intelligent Systems, 21*, 70–73.

Klein, G., Phillips, J. K., Rall, E. L. and Peluso, D. A. 2007. A data-frame theory of sensemaking. In R. Hoffman (Ed.), *Expertise Out of Context: Proceedings of the 6th International Conference on Naturalistic Decision Making*, Mahwah, NJ: Erlbaum, 113–155.

Klein, G. A. 1993. A recognition-primed decision (RPD) model of rapid decision making. In G. A. Klein, J. Orasanu, R. Calderwood and C. E. Zsambok (Eds.), *Decision Making in Action: Models and Methods*, Norwood, NJ: Ablex, 138–147.

Landgren, J. 2004. Fire crew enroute sensemaking in emergency response. *Proceedings of the 1st International Conference on Information Systems for Crisis Response and Management (ISCRAM2004)*, 3–4 May, Brussels, Belgium.

Landgren, J. 2005a. Shared use of information technology in emergency response work: Results from a field experiment. *Proceedings of the 2nd Conference on Information Systems for Crisis Response and Management (ISCRAM2005)*, 18–20 April, Brussels, Belgium, 1–7.

Landgren, J. 2005b. Supporting fire crew sensemaking en route to incidents. *International Journal of Emergency Management*, 2, 176–188.

Landgren, J. 2007. A study of emergency response work: Patterns of mobile phone interaction. *In Proceedings of the 2007 SIGCHI Conference on Human Factors in Computing Systems (CHI'2007)*, San Jose, CA, 1323–1332.

Larkin, K., McDermott, J., Simon, D. and Simon, H. 1980. Expert and novice performance in solving physics problems. *Science*, 208, 1335–1342.

Leedom, D. K. 2001. *Final Report: Sensemaking Symposium*. 23–25 October. Available at http://www.dodccrp.org/files/sensemaking_final_report.pdf.

Lipshitz, R. 1993. Converging themes in the study of decision making in realistic settings. In G. A. Klein, J. Orasanu, R. Calderwood and C. E. Zsambok (Eds.), *Decision Making in Action: Models and Methods*, Norwood, NJ: Ablex, 103–137.

Maglio, P. P., Matlock, T., Raphaely, D., Chernicky, B. and Kirsh, D. 1999. Interactive skill in Scrabble. *Proceedings of the Twenty-First Annual Conference of the Cognitive Science Society*, Englewood Cliffs, NJ: Lawrence Erlbaum Associates, 326–330.

Maitlis, S. and Christianson, M. 2014. Sensemaking in organizations: Taking stock and moving forward. *The Academy of Management Annals*, 8, 57–125.

Manning, P. 1988. *Symbolic Communication: Signifying Calls and the Police Response*. Cambridge, MA: MIT Press.

McMaster, R. and Baber, C. 2005. Integrating human factors into systems engineering through a distributed cognition notation. *IEE People and Systems Symposium: Who Are We Designing For?* 16–17 November, London: Institute of Engineering Technology, 77–83.

McMaster, R., Baber, C. and Houghton, R. J. 2006. Investigating Distributed Cognition Processes during Effects-Based Operations, Proceedings of The Technical Cooperation Program Human Systems Integration Symposium. Yeovil: Human Factors Integration Defence Technology Centre, 4–5.

McMaster, R. and Baber, C. 2006. Assessing the impact of digital communications technology on existing C2 systems; a distributed cognition perspective. *Proceedings of the 11th International Command and Control Research and Technology Symposium: Coalition Command and Control in the Networked Era*, Washington, DC: CCRP, 26–28.

McMaster, R. and Baber, C. 2012. Multi-agency operations: Cooperation during flooding. *Applied Ergonomics*, 43, 38–47.

McNeese, M. D., Pfaff, M. S., Connors, E. S. and Obieta, J. F. 2006. Multiple vantage points of the common operational picture: Supporting international teamwork. *Proceedings of the Human Factors and Ergonomics Society 50th Annual Meeting*, Santa Monica, CA: Human Factors and Ergonomics Society, 467–471.

Mendonça, D., Jefferson, T. and Harrald, J. 2007. Collaborative adhocracies and mix-and-match technologies in emergency management. *Communications of the ACM, 50*, 44–49.

MoD. 2005. *Network Enabled Capability*. JSP 777 Edn1. Ministry of Defence UK.

Mohammed, S., Ferzandi, L. and Hamilton, K. 2010. Metaphor no more: A 15-year review of the team mental model construct. *Journal of Management, 36*, 876–910.

Nemeth, C. and Cook, R. 2004. Discovering and supporting temporal cognition in complex environments. In *Proceedings of the National Conference of the Cognitive Science Society*, August, Chicago, 1005–1010.

Norman, D. A. 1991. Cognitive artifacts. In J. M. Carroll (Ed.), *Designing Interaction: Psychology at the Human–Computer Interface*, Cambridge, UK: Cambridge University Press.

Norman, D. A. 1993. *Things that Make Us Smart*. Menlo Park, CA: Addison-Wesley.

Norman, D. A. and Draper, S. W. 1986. *User Centered System Design*, Boca Raton, FL: CRC Press.

NPIA. 2009. *Guidance on multi-agency interoperability*. London: National Police Improvement Agency. Retrieved May 2012 from: http://www.acpo.police.uk/documents/uniformed/2009/200907UNMAI01.pdf.

NPIA. 2010. *Standard operating procedure guide on multi-agency Airwave interoperability*. London: National Police Improvement Agency. Retrieved November 2012 from: http://www.npia.police.uk/en/docs/SOP_Guide_on_Multi-Agency_Airwave_Interoperability_Feb2011.pdf.

O'Connor, L. 2010. Workarounds in accident and emergency and intensive therapy departments: Resilience, creation and consequences. MSc thesis, University College London.

Ormerod, T., Barrett, E. and Taylor, P. J. 2008. Investigative sense-making in criminal contexts. In J. M. Schraagen (Ed.), *Naturalistic Decision Making and Macrocognition*, Avebury: Ashgate, 81–102.

Paley, M. J., Linegang, M. P. and Morley, R. M. 2002. Using communication data to assess organizational and system effectiveness in future combat systems. In *Proceedings of the Human Factors and Ergonomics Society Annual Meeting*, Santa Monica, CA: Human Factors and Ergomoics Society, 290–294.

Parker, S. K. and Wall, T. D. 1998. *Job and Work Design*. London: Sage.

Perry, M. 2003. Distributed cognition. In J. M. Carroll (Ed.), *HCI Models, Theories and Frameworks: Towards a Multidisciplinary Science*. London: Morgan Kaufmann.

Perry, M. 2013. Socially distributed cognition in loosely coupled systems. In S. J. Cowley and F. Vallée-Tourangeau (Eds.), *Cognition Beyond the Brain: Computation, Interactivity and Human Artifice*, London: Springer, 147–169.

Pfaff, M. S., Drury, J. L., Klein, G. L., More, L., Moon, S. P. and Liu, Y. 2010. Weighing decisions: Aiding emergency response decision making via option awareness. In *2010 IEEE International Conference on Technologies for Homeland Security (HST)*, New York: IEEE, 251–257.

Pirolli, P. and Card, S. 2005. The sensemaking process and leverage points for analyst technology as identified through cognitive task analysis. In *Proceedings of the International Conference on Intelligence Analysis*, 2–4 May, McLean, VA, 1–6.

Pirolli, P. and Russell, D. M. 2011. Introduction to this special issue in sensemaking. *Human–Computer Interaction, 26*, 1–8.

Popp, R. and Poindexter, J. 2006. Countering terrorism through information and privacy protection technologies. *IEEE Security & Privacy, 4*, 18–27.

Potter, S., Kalfoglou, Y., Alani, H., Bachler, M., Buckingham Shum, S., Carvalho, R., Chakravarthy, A., Chalmers, S., Chapman, S., Hu, B., Preece, A., Shadbolt, N., Tate, A. and Tuffield, M. 2007. The application of advanced knowledge technologies for emergency response, 4th International Information Systems for Crisis Response and Management (ISCRAM 2007), Delft, The Netherlands.

Quarantelli, E. L. 1997. Problematical aspects of the information/communication revolution for disaster planning and research: Ten non-technical issues and questions. *Disaster Prevention and Management, 6*, 94–106.

Rahman, A. 1996. Peoples' perception and response to floodings: The Bangladesh experience. *Journal of Contingencies and Crisis Management, 4*, 198–207.

Ramduny-Ellis, D., Dix, A., Rayson, P., Onditi, V. Sommerville, I. and Ransom, J. 2005. Artefacts as designed, artefacts as used: Resources for uncovering activity dynamics. *Cognition, Technology and Work, 7*, 76–87.

Reddy, M., Paul, S. A., Abraham, J., McNeese, M. D., deFlitch, C. J., and Yen, J. 2009. Challenges to effective crisis management: Using information and communication tools to coordinate emergency medical services and emergency department teams. *International Journal of Medical Informatics, 78* (4), 259–269.

Roberts, K. H. and Rousseau, D. M. 1989. Research in nearly failure-free, high-reliability organizations: Having the bubble. *IEEE Transactions on Engineering Management, 36*, 132–139.

Rochlin, G. I. 1991. Trapped in the Net: The unanticipated consequences of computerization. Princeton, NJ: Princeton University Press.

Rogers, Y. and Ellis, J. 1994. Distributed cognition: An alternative framework for analysing and explaining collaborative working. *Journal of Information Technology, 9*, 119–128.

Rojek, J. and Smith, M. R. 2007. Law enforcement lessons learned from Hurricane Katrina. *Review of Policy Research, 24*, 589–608.

Scaife, M. and Rogers, Y. 1996. External cognition: How do graphical representations work? *International Journal of Human–Computer Studies, 45*, 185–213.

Schneider, S. K. 2005. Administrative breakdowns in the governmental response to Hurricane Katrina. *Public Administration Review, 65*, 515–516.

Schraagen, J. M., Klein, G. and Hoffman, R. R. 2008. The macrocognition framework of naturalistic decision making. In J. M. Schraagen, L. Militello, T. Omerod and R. Lipshitz (Eds.), *Naturalistic Decision Making and Macrocognition*, Aldershot, UK: Ashgate, 3–25.

Schwartz, S. and Griffin, T. 1986. *Medical Thinking: The Psychology of Medical Judgement and Decision Making*. New York: Springer.

Seagull, F. J., Plasters, C., Xiao, Y. and Mackenzie, C. F. 2003. Collaborative management of complex coordination systems: Operating room schedule coordination. *Proceedings of the Human Factors and Ergonomics Society Annual Meeting*, Santa Monica, CA: HFES, 1521–1525.

Serfaty, D., Entin, E. E. and Volpe, C. 1993. Adaptation to stress in team decision-making and coordination. In *Proceedings of the Human Factors and Ergonomics Society Annual Meeting*, Santa Monica, CA: Human Factors and Ergonomics Society, 1228–1232.

Simon, H. A. 1956. Rational choice and the structure of the environment. *Psychological Review, 63*, 129–138.

Smith, W. and Dowell, J. 2000. A case study of co-ordinative decision-making in disaster management. *Ergonomics, 43*, 1153–1166.

Snow, J. and Manning, L., 2007. Saving Walham. Channel 4 News. Available at http://www.channel4.com/news/articles/society/environment/saving þwalham/624157#fold (accessed January 2009).

Stanton, N., Baber, C. and Harris, D. 2008. *Modelling Command and Control*. Aldershot, UK: Ashgate Publishing.

Stanton, N. A., Stewart, R., Harris, D., Houghton, R. J., Baber, C., McMaster, R., Salmon, P. et al. 2006. Distributed situation awareness in dynamic systems: Theoretical development and application of an ergonomics methodology. *Ergonomics, 49*, 1288–1311.

Stewart, R. J., Stanton, N. A., Harris, D., Baber, C., Salmon, P., Mock, M., Tatlock, K., Wells, L. and Kay, A. 2008. Distributed situational awareness in an airborne warning and control aircraft: Application of a novel ergonomics methodology. *Cognition, Technology and Work, 10*, 221–229.

Stone, N. J. and Posey, M. 2008. Understanding coordination in computer-mediated versus face-to-face groups. *Computers in Human Behavior, 24*, 827–851.

Suchman, L. A. 1987. *Plans and Situated Actions: The Problem of Human-Machine Communication*. Cambridge, MA: Cambridge University Press.

Taylor, J. R. and van Every, E. J. 1999. *The Emergent Organization: Communication as Its Site and Surface*. Mahwah, NJ: LEA.

Taylor, S. E. and Crocker, J. 1981. Schematic bases of social information processing. In *Social Cognition: The Ontario Symposium, 1*, 89–134.

Thompson, L. and Fine, G. A. 1999. Socially shared cognition, affect, and behavior: A review and integration. *Personality and Social Psychology Review, 3* (4), 278–302.

Tolk, A. 2003. Beyond technical interoperability-introducing a reference model for measures of merit for coalition interoperability. Norfolk, VA: Old Dominion University.

Townsend, M. 2007. Why won't the US tell us how Matty died. The Observer, Sunday 4th February 2007. Available at http://www.theguardian.com/uk/2007/feb/04/iraq.military.

Turner, P. 2007. The role of sensemaking in the Command-ISTAR relationship. *12th International Command and Control Research and Technology Symposium: Adapting C2 to the 21st Century*, 19–21 June, Newport, RI.

Tversky, A. and Kahneman, D. 1974. Judgement under uncertainty: Heuristics and biases. *Science, 185*, 1124–1131.

Umapathy, K. 2010. Requirements to support collaborative sensemaking. In *Proceedings of the International Workshop on Collaborative Information Seeking*, Savannah, GA, United States, February.

Vallée-Tourangeau, F. and Cowley, S. J. 2013. Human thinking beyond the brain. In S. J. Cowley and F. Vallée-Tourangeau (Eds.), *Cognition Beyond the Brain: Computation, Interactivity and Human Artifice*, London: Springer, 1–11.

Van Fenema, P. C. 2005. Collaborative elasticity and breakdowns in high reliability organizations: Contributions from distributed cognition and collective mind theory. *Cognition, Technology & Work, 7*, 134–140.

von Lubitz, D. K. J. E., Beakley, J. E. and Patricelli, F. 2008. 'All hazards approach' to disaster management: The role of information and knowledge management, Boyd's OODA loop, and network-centrality. *Disasters, 32*, 561–585.

Vygotsky, L. S. 1978. *Mind in Society: The Development of Higher Psychological Processes*. Cambridge, MA: Harvard University Press.

Walker, G. H., Stanton, N. A., Stewart, R., Jenkins, D., Wells, L., Salmon, P. and Baber, C. 2009. Using an integrated methods approach to analyse the emergent properties of military command and control. *Applied Ergonomics, 40*, 636–647.

Warwickshire Police. 2005. *National Call Handling Standards*, Internal Training Document.

Whalen, M. R. and Zimmerman, D. H. 1990. Describing trouble: Practical epistemology in citizen calls to the police. *Language in Society, 19*(4), 465–492.

Weick, K. 1995. *Sensemaking in Organizations*. Thousand Oaks, CA: Sage.

Weick, K. E. 1988. Enacted sensemaking in crisis situations. *Journal of Management Studies, 25*, 305–317.

Weick, K. E. Sutcliffe, K. M. and Obstfeld, D. 2005. Organizing and the process of sensemaking. *Organization Science, 16*, 409–421.

Wenger, E. 1998. *Communities of Practice: Learning, Meaning and Identity*. Cambridge, UK: Cambridge University Press.

Wenger, E., McDermott, R. and Snyder, W. M. 2002. *Cultivating Communities of Practice: A Guide to Managing Knowledge*. Boston: Harvard Business School Press.

Westrum, R. 1982. Social intelligence about hidden errors. *Knowledge, 3*, 381–400.

Wolbers, J. and Boersma, K. 2013. The common operational picture as collective sensemaking. *Journal of Contingencies and Crisis Management, 21*, 186–199.

Wright, P. C., Fields, B. and Harrison, M. D. 1996. Distributed information resources: A new approach to interaction modelling. *Proceedings of ECCE-8: The 8th European Conference on Cognitive Ergonomics*, 10–13 September, Roquencourt, France: European Association of Cognitive Ergonomics, 5–10.

Wright, P. C., Fields, R. E. and Harrison, M. D. 2000. Analyzing human–computer interaction as distributed cognition: The resources model. *Human–Computer Interaction, 15*, 1–41.

Wright, P. C., Pocock, S. and Fields, R. E. 1998. The prescription and practice of work on the flight deck. In T. R. G. Green, L. Bannon, C. P. Warren and J. Buckley (Eds.), *ECCE9, Ninth European Conference on Cognitive Ergonomics*, Limerick, Ireland, EACE, 37–42.

Wu, A., Convertino, G., Ganoe, C., Carroll, J. M. and Zhang, X. L. 2013. Supporting collaborative sense-making in emergency management through geo-visualization. *International Journal of Human-Computer Studies, 71*, 4–23.

Zhang, J. and Norman, D. A. 1994. Representations in distributed cognitive tasks. *Cognitive Science, 18*, 87–122.

Appendix A: Timelines of Events in London Bombings (2005)

LOCATION	0850	0855	0900	0905	0910	0915	0920	0925	0930	0935	0945	0950	0955
Aldgate	**0853**: Bomb explodes between Aldgate and Liverpool Street stations **0855**: BTP report 'loud bang' at Liverpool Street	**0856**: Fire called to Aldgate **0858**: BTP identify Aldgate as site of incident **0859**: LU control call all services to Edgware Road, Aldgate, King's Cross	**0900**: Fire at Aldgate **0901**: BTP request ambulance to Aldgate	**0902**: Senior fire on scene **0903**: Ambulance at Liverpool Street **0905**: Fire declare major incident **0907**: BTP report 25 walking wounded **0907**: Ambulance alert hospitals and declare major incident **0908**: BTP report train accident and declare major incident	**0910**: City of London Police report explosion and declare major incident **0914**: Ambulance at Aldgate and report bomb and 5 fatalities	**0919**: BTP request assistance from Met Police	**0920**: Met Police at Aldgate						

(Continued)

LOCATION	0850	0855	0900	0905	0910	0915	0920	0925	0930	0935	0945	0950	0955
King's Cross/Russell Square	**0853**: Bomb explodes between King's Cross and Russell Square stations	**0856**: Met Police CCTV note incident at King's Cross **0859**: LU control call all services to Edgware Road, Aldgate, King's Cross	**0902**: Fire called to 'smoke coming out of tunnel' at King's Cross **0904**: Ambulance to Euston Square (3) and to King's Cross (1)	**0907**: Fire to Euston Square	**0913**: Ambulance to King's Cross	**0915**: Met Police to King's Cross **0918**: Ambulance called to Russell Square **0919**: Fire to King's Cross (5)				**0930**: Ambulance to Russell Square **0938**: Ambulance declare major incident at Russell Square			
Edgware Road	**0851**: Bomb explodes between Edgware Road and Paddington	**0858**: Public call from Praed Street **0858**: BTP called to incident at Edgware Road **0859**: LU control call all services to Edgware Road, Aldgate, King's Cross	**0904**: Fire to Praed Street (5) **0904**: Fire contact Met Police	**0907**: Fire control receive request to attend Edgware Road	**0912**: Met Police at Edgware Road **0912**: Ambulance at Edgware Road **0914**: Ambulance report explosion and 1000 casualties	**0916**: Ambulance request back-up **0918**: Fire at Edgware Road	**0924**: Ambulance declare major incident		**0932**: Met Police at Edgware Road **0934**: Fire at Edgware Road	**0937**: Fire deployed from Praed Street to Edgware Road			

(Continued)

LOCATION	0850	0855	0900	0905	0910	0915	0920	0925	0930	0935	0945	0950	0955
Tavistock Square											**0947:** Bomb explodes on a bus in Tavistock Square **0947:** Met Police officer at the scene reports the explosion **0948:** First call from member of the public	**0950:** Fire Brigade attend	**0957:** Ambulance attends (arrives en route to a different incident)

Appendix B: Social Network Analysis Centrality Scores

AGENT	DEGREE CENTRALITY	DISTANCE CENTRALITY
London Fire Brigade (control)	**0.789**	**29.96**
London Ambulance Service (control)	**0.526**	48.55
British Transport Police (BTP; control)	**0.228**	**31.29**
Metropolitan Police Service (control)	**0.158**	**17.38**
BTP unit 1	**0.105**	108.31
Ambulance 3	**0.088**	48.55
London Underground Network Control Centre	0.088	**12.92**
Ambulance 1	**0.070**	48.55
Fire 1	**0.070**	**21.66**
Fire 18	**0.070**	**21.66**
Ambulance 4	0.053	48.55
Planning manager 1	0.053	48.55
Fire 10	0.053	**21.66**
Ambulance 2	0.035	48.55
Fast response unit 1	0.035	48.55
Fast response unit 2	0.035	48.55
Fire 9	0.035	**21.66**
Fire 15	0.035	**21.66**
Fire 16	0.035	**21.66**
Fire rescue unit 2	0.035	**21.66**
Met unit 2	0.035	108.31
Met unit 3	0.035	108.31

(*Continued*)

AGENT	DEGREE CENTRALITY	DISTANCE CENTRALITY
BTP unit 2	0.018	108.31
City of London Police (control)	0.018	156.44
Ambulance 5	0.018	48.55
Professional standards officer	0.018	48.55
Fire 2	0.018	**21.66**
Fire 3	0.018	**21.66**
Fire 4	0.018	**21.66**
Fire 5	0.018	**21.66**
Fire 6	0.018	**21.66**
Fire 7	0.018	**21.66**
Fire 8	0.018	**21.66**
Fire 11	0.018	**21.66**
Fire 12	0.018	**21.66**
Fire 13	0.018	**21.66**
Fire 14	0.018	**21.66**
Fire 17	0.018	**21.66**
Fire 19	0.018	**21.66**
Fire 20	0.018	**21.66**
Fire investigation unit 1	0.018	**21.66**
Fire rescue unit 1	0.018	**21.66**
Senior fire officer	0.018	**21.66**
All other fire units (East London)	0.018	**21.66**
London Underground staff (Aldgate)	0.018	108.31
Met unit 1	0.018	108.31
Met unit 4	0.018	108.31
Member of public 1	0.018	**21.66**
Member of public 2	0.018	**21.66**
Member of public 3	0.018	48.55
Member of public 4	0.018	48.55
Member of public 5	0.018	**21.66**
Unknown 1	0.018	48.55
Unknown 2	0.018	**21.66**
Unknown 3	0.018	108.31
Unknown 4	0.018	48.55
Unknown 5	0.018	**21.66**
Unknown 6	0.018	**21.66**

Index

This index includes preface and appendices. Page numbers with f, n, and t refer to figures, footnotes, and tables, respectively.

A

Abbreviations, 106
Abductive reasoning, 15
ACC, *see* Assistant chief constable (ACC)
Access overload control (ACCOLC), 184
ACCOLC, *see* Access overload control (ACCOLC)
ACPO, *see* Association of Chief Police Officers (ACPO)
Acronyms, 106
Actionable knowledge, 202
Active listening, 110, 132
Activities, 18t, 40
Acute mental health problems, 102
Adams, F., 53, 55
Adaptation, 58
Ad hoc groups, 87, 94, 133, 139, 172, 220, 221–222

Agencies
  multi-
    emergency response, 140–141, 140t
    major incident doctrine, 162
    response of, earthquake in California, 190
Airspeed indicator, 44, 45f
Airwave system
  digital radio network, 171, 208–209, 213
  radios, 170
  user groups, 170
Aizawa, K., 53, 55
Alberts, D. S., 78, 79
Aldgate Station, 175f, 184, 185f, 242
Ambulance Control Centre, 80f
Ambulance trusts, 75
Appropriated artefacts, 47–48, 48f
Arson, 101

247

INDEX

Artefact-driven sensemaking; *see also* Artefacts
   artefacts
      external representations, 42–43
      part of a cognitive system, 43–45
      resources for action, 45–49
   conclusions, 55–56
   definition, 13–14
   distributed cognition and the extended mind, 52–55
   distribution, 49
   introduction, 39–42
   reflections, 216–217
   representation construction, 49–52, 50f
Artefacts
   appropriated, 47–48, 48f
   call handlers, 122t, 123
   cognitive, 39, 58, 129, 148
   controllers, 122t
   cycle, task-, 12, 12f
   design of, 45
   distribution of, 49
   as external media, 57
   external representations, 40, 42–43
   formal, 42
   improvised, 131–132
   informal, 42
   low-tech, 42, 43
   making sense through, 129–131, 161–162
   making sense with, 127–129, 160–161
   modifications, 104
   part of a cognitive system, 43–45
   police incident response sensemaking, 128t
   representational, 91t
   resources for action, 45–49
   responding officers, 122t
   role in distributed SA, 69
   routine incident response, 122t
   sensemaking, *see* Artefact-driven sensemaking
   shared, 64
   used for communications, 133
   as used, 47
Ascoli, Giorgio A., 21
Assaults, 102
Assistant chief constable (ACC), 77
Association of Chief Police Officers (ACPO), 172, 173
Avon Fire and Rescue Service, 82, 143, 144f, 145
Awareness, 97, 140t, 153; *see also* Situation awareness (SA)

B

Baber, Christopher, 9, 17, 42, 48, 129
Bangladesh flood (1988), 140
Baron, R. M., 23
Biases, 31–36
   frames and, 32–34
   overview, 31–32
   sources
      in sensemaking, 34–36
      typical, 31
Blackboard architecture, 193
Blandford, A., 51, 65
Boersma, K., 200, 202
Bolstad, C. A., 66, 67
Boston Marathon Bombing (2013), 176–178, 177f, 182
Boyd cycle, 82, 83f, 84
Breivik, Anders, 1
British Transport Police (BTP), 80f, 175f, 185f, 187, 188, 245
Bronze (operational) command, 74–75, 74f, 78f, 144f; *see also* Walham Floods (2007)

INDEX

BTP, *see* British Transport Police (BTP)
Burglaries, in progress, 101, 111f
Burnett, M., 92, 163

C

C2, *see* Command and control (C2)
California, 190
Caller, 103, 104, 105, 108, 122f, 130
Call handlers
  active listening, 110
  inflexibility of incident log, 123
  interpreting problem, 130
  notepad, 121, 122f, 122t, 123, 128t
  routine incident response, 74, 103f, 104–106, 108, 122t
  sensemaking process, 101, 121, 122f
  workstation, 105f
Call process
  abbreviations and acronyms, 106
  callers, 104–105
  call handlers, 104–106, 105f
  dispatcher, 104–105
  open questions, 105–106
  process flow, 107f
  sensemaking in, 121, 122f, 123
Card, S., 49
Carroll, John M., 12
Case Studies
  Boston Marathon Bombing (2013), 176–178, 177f
  London Bombing (2005), 173–176, 174f, 175f
  Walham Floods (2007), 141–143, 142f
CAST, *see* Co-ordinated awareness of situations by teams (CAST)
Casualties, Hazards, Access, Location, Emergency Services, Type (CHALET), 197, 198f
Category 1 responders, 75, 157

Centrality
  degree, 187
  distance, 188
  fallacy of, 94, 97–98
  social network analysis scores, 245–246
Central Motorway Police Group, 80f
Chains of command, poor, 140t
CHALET, *see* Casualties, Hazards, Access, Location, Emergency Services, Type (CHALET)
Chalmers, D. J., 53
City of London Police, 175f, 184, 185f, 246
Civil Contingencies Act (2007), 157
Clark, Andy, 53, 54
Clark, Herbert, 9, 10
CMC, *see* Computer-mediated communication (CMC)
Cognition, systems-level, 41, 58
Cognitive actions, 39, 53
Cognitive artefacts
  concept, 148
  language as, 58
  significance, 39
  police officers, 116
  private, 58, 121, 129
  private shared, 118
  public, 129
Collaboration, 47, 132–133, 162–165
Collaborative networks
  community of practice and exploration network, 92–93, 93t
  tightly and loosely coupled work systems, 91, 91t
  traditional hierarchical C2 and edge networks, 92t, 92
Collaborative search
  enactment, 60–61
  extracted cues, 62
  identity, 59–60

after meaning, 59–64
ongoing, 61–62
plausible rather than true, 62–64
retrospective, 60
social, 61
Collaborative sensemaking; *see also* Common operating procedure (COP)
collaborative search after meaning, 59–64
definition, 14
digital radio and, 128t, 208
distributed cognition in major incidents, 162–165
emergency medicine in, 42
geo-visualization tool, 203, 205
problem of situation awareness, 64–69
reflections, 217–219
routine incident response and, 135
sensemaking as system activity, 70–71, 70f
social processes and, 163–165
socially shared cognition, 57–59
Command, definition, 78
Command and control (C2)
constraint, 17
definition, 78
emergency response, 73–74
bronze (operational), 74–75, 74f, 78f
category 1 response agencies, 75
gold (strategic), 74f, 75, 76, 78f
silver (tactical), 74f, 75, 76, 78f
essential capabilities, 78–79
future of, 86–87
large-scale disasters, 43
network, 79, 80f
network agents, 103
police incident response, 79, 80f, 81f, 81

purpose, 73
responding units, 59
revisiting, 221–222
sensemaking
big-picture view operations, 89–91
collaborative networks, 91–93
co-ordination, 98–99
planning and adaptation, 93–96
as prerequisite, 79
problem detection, 96–98
structures, 140t
traditional hierarchical, 92t, 92, 205
UK emergency services, 73–87
Commanders, 27, 221n; *see also* Bronze (operational) command; Gold (strategic) command; Silver (tactical) command
Committed action, 23, 202
Committed interpretation, 23, 202
Common ground
in communications, 8–12, 17
community evidence, 10, 11
definition, 9
distributed cognition in major incidents, 155–166
linguistic evidence, 10, 11
misuse, 66
perceptual evidence, 10, 10–11
sensemaking and, 7, 11–12
significance, 64
Common operating picture (COP), 193–210
blackboard architecture and, 193
conclusions, 210
informing versus understanding, 194–196
ontologies, 197–203
as product or process, 206–210
significance, 194

situation space versus decision space, 206
state of the world representation or collaborative planning tool, 203–206
Common reference, communication, 47
Common relevant operating picture (CROP), 197
Common representations, 58, 59
Common vocabulary, 77, 191
Communications
  C2 network, 79, 91
  collaborative sensemaking, 58
  command centres, 79, 80f
  common ground, 8–12
  common reference, 47
  common vocabulary, 191
  controllers, co-ordination example, 108–110
  electronic, 93–94
  equipment, 151
  failure, 140, 151–152, 164, 184
  incident response, 103f
  inter-agency, 140t
  lack of, control rooms, 186
  non-urgent, 110
  officers and controllers, 128
  talk group, 112–113
  verbal, 64
Communications equipment, 151
Community evidence, 10, 11
Community of practice, 92–93, 93t
Competitive practices, 140t
Complex learning, 15f
Computer-mediated communication (CMC), 98
Consolidation, 138
Control, 78, 104
Controllers
  artefacts, 111, 122t
  closure of incident, 120
  co-ordination example, 108–110
  formalised language, 123
  frame-defined data collection, 110–112
  framing incident management problems, 126–127
  incident details, 108
  incident logs and, 130
  local, 108, 110, 133
  open incident list, 124f
  police officer's enquiries, 115
  primary resource management tool, 126
  resource allocation to incidents, 123–126
  routine incident response, 122t
  traffic, 103f, 110, 122t
  Warwickshire Police, 110n
  workstation, 108, 108f, 124f
Control rooms, 110n, 122t, 186
Conversations, common ground, 8–12
Co-operation, 139, 151–152
Co-ordinated awareness of situations by teams (CAST), 69
Coordination, 140t
Co-ordination, 15f, 58, 98–99, 140, 152
COP, *see* common operating picture (COP)
Coupling-constitution fallacy, 54
Course of action (COA), optimal, 206
Crime scene investigation as distributed cognition process, 48
Crime Survey for England and Wales, 34
Criminal damage, 101
Critical decision method (CDM) questions, various agency, 156t

252 INDEX

Crocker, J., 24
CROP, *see* common relevant operating picture (CROP)
Cultural heritage, 58, 217

D

Data
  elements, 15
  focusing, 51
  information relevance and, 192
  information semantics and, 191
  quality, 191–192
  quantity, 191–192
Data-based frame modification, 29
Data elements, 15
Data focusing, 51
Data-frame model, 27–30, 29f, 63
Data quality, 191–192
Data quantity, 191–192
Data terminal, 209, 209f
DCFO, *see* deputy chief fire officer (DCFO)
Decision inertia, 191
Decision making
  natural, 15f
  sensemaking and, 26–27
Degree centrality, 187, 245–246
De Jaegher, H., 61
Delays, 59, 175, 186
Deputy chief fire officer (DCFO), 143, 144f, 145
Detecting problems, 15f
DHS, *see* US Department for Homeland Security
Digital radio
  as artefact, 104, 116, 118, 122t, 128t
  in emergency response, 80f
  formal purpose, 48, 79, 128t
  police incident response sensemaking, 128t
  sensemaking role, 128t, 208, 224

Di Paolo, E., 61
Dispatcher, 4, 6, 11, 104–105
Disposable gloves, 131
Distributed cognition
Distance centrality, 188
Distributed cognition
  cognitive processes, 41
  crime scene investigation as, 48
  extended mind and, 52–55
    mark of the cognitive, 53–55
    Otto's notebook, 52–53
  features of activities in task, 18t
  graphical representation, 17–20
  major incidents, 151–166
    common ground, 155–160
    conclusions, 165–166
    introduction, 151–155
    making sense through artefacts, 161–162
    making sense through collaboration, 162–165
    making sense with artefacts, 160–161
  military exercise, 17–18
  overview, 16–17
  routine incidents, 121–135
    allocating resources to incidents, 123–127
    call process, 121, 122f, 123
    conclusions, 134–135
    improvised artefacts, 131–132
    making sense through artefacts, 129–131
    making sense through collaboration, 132–133
    making sense with artefacts, 127–129
    routine incident response, 121, 122t
  sensemaking as, 219–221
  target review process, 18, 19f, 20
Domestic violence, 101
Drury, J. L., 206

Dubrovsky, V. J., 93
Dutch emergency management system, 202

E

EA, *see* Environment Agency (EA)
East London Fire Units, 185f
Edge networks, 92t, 92, 205
Edgware Road, 175, 175f, 243
Electronic communications, 93–94
Emergency dispatch, 110
Emergency medicine, 42, 65
Emergency Operations Centre (EOC), 95
Emergency response
    analysing network structures and interoperability, 184–189
    characteristics, 5
    challenge of sharing information, 190–192
        data and information relevance, 192
        data and information semantics, 191
        data quality, quantity and timeliness, 191–192
    command and control, 73–74
        bronze (operational), 74–75, 74f, 78f
        category 1 response agencies, 75
        gold (strategic), 74f, 75, 76, 78f
        silver (tactical), 74f, 75, 76, 78f
    commanders in, 221n
    delays, 186
    interoperability, 225
    delays, 186
    sensemaking as a social process, 182–184
    as source of information, 197
Emergency response centre (ERC), 202–203
Emergency services, *see* Emergency response
Endsley, M. R., 66, 67
England, 75, 142
Enquiries, officer-initiated database, 110
Environment Agency (EA),
    communication media, 151
    critical decision method (CDM) questions, 156t
    legislations, 157
Environmental perception, 59–60
EOC, *see* Emergency Operations Centre (EOC)
Evidence file, 50f, 50
Experienced fire service commanders, 24
Expertise, subtle cues, 96
Experts (sense makers), 7, 16; *see also* Individual sensemaking
Exploration networks, 92–93, 93t, 163–164
External data sources, 50f, 50
External representations, 40, 42–43, 50, 51, 127

F

Face-to-face (FtF), 93, 98, 191, 202
Faisal, S., 52
Fallacy of centrality, 94, 97–98
FCC, *see* Force Communications Centre (FCC)
Fire and rescue services, 73n, 75, 157
Fire Brigade Call Centre, 80f
Flexecution, 30–31
Flexible execution, 30–31
Flows, 71
Force Communications Centre (FCC), 79, 103, 122t
Form, 18t

INDEX

Formal artefacts, 42
Fraher, A. L., 91
Frame construction/modification, 28
Frame-defined data collection
    definition, 28, 29, 29f
    by call handler's, 110–112, 117f, 122f, 130
    police incident response sensemaking, 110–112, 117f, 128t
Frames; *see also* Data-frame theory
    as external representations, 40, 52
    as internal representations, 40
    as sensemaking characteristics, 15, 16
    biases and, 32–34
    definition, 28n, 29f
    distributed SA and, 65, 67, 69
    incident, 164
Frame-seeking data collection, 117f, 122f, 130
Frame testing, 122f

G

General announcements, 110
Geographical information system (GIS), 110, 123, 124f, 125–126, 195, 203
Geo-visualization tool, 203, 205
Gigerenzer, G., 33
Gioia, D. A., 63
GIS, *see* Geographical information system (GIS)
Gjørv Commission, 1, 2, 4
Gloucestershire, 141–143
Gloucestershire Fire and Rescue service, 143, 144, 144f, 145, 151–156t
Gloucestershire Police, 73n
Gloves, disposable, 131

Gold (strategic) command; *see also* Walham Floods (2007)
    Boston Marathon Bombings (2013), 176
    function, 75, 76, 78f
    hierarchical structure, 74f, 78f
    national decision model (NDM), 81, 81f
    police, 77
Gold liaison, 143, 144f
Gorman, J. C., 66, 69
Guidance on Multi-agency Interoperability (2009), 194

H

Hall, D. L., 206
Handling, 105
Harrald, J., 190
Hayes, R. E., 78, 79
Hazards, 85f
Heath, C., 64
HFI-DTC, *see* Human Factors Integrated Defence Technology Centre (HFI-DTC)
Houghton, R. J., 187
Human Factors Integrated Defence Technology Centre (HFI-DTC), xix
Hurricane Katrina, 140
Hutchins, E., 44, 69
Hypotheses, 50, 50f

I

ICU, *see* Incident command unit (ICU)
Identity, 59–60
Improvised artefacts, 131–132
IMS, *see* Incident management system (IMS)

Incident command unit (ICU), 145–149, 145f
Incident details, 108, 116
Incident frame, 164
Incident logs, 123, 125, 130, 203, 204f
Incident management system (IMS)
 as artefact, 104, 122t, 160
 function, 79, 103f, 208
 issues, 113–114
 log, 108–110, 128t
 open incident list, 123, 124f, 125, 126
 pocket notebook versus, 120
 significance, 110, 111f
 target review process, 19f
Incident response, 222–225
Incidents
 definition, 138
 resource allocation
  controllers, 123–126
  police officers, 125, 126–127
Indian Ocean Tsunami (2004), 140
Individual commander, 27
Individual sensemaking
 cognition, 21–23
 data-frame model, 27–30
 definition, 13
 flexecution, 30–31
 prior experience representation, 23–27
 problem of bias, 31–36
 reflections, 215–216
Inferences, 15
Informal artefacts, 42
Informal sense, 62
Information
 ambiguous, 59
 bottlenecks, 205
 challenge of sharing, 190–192
  data and information relevance, 192
  data and information semantics, 191
  data quality, quantity and timeliness, 191–192
 class diagram, 199f, 200
 as cognition in the human mind, 68
 crowdsourced, 197
 extracted cues, 62
 failure for dissemination, 140t
 failure to share, 140
 foraging, activities, 50
 recognition, 2
 relevance, 192
 representation, 200
 semantics, 191
 shared displays, 194
 sources, 10, 197
 stovepiping, 186, 205
 structured, 200
 triage, 191
 user interfaces, 200, 201f
Information bottlenecks, 205
Information displays, shared, 194
Information foraging, activities, 50–51
Information overload, 205
Information relevance, 192
Information semantics, 191
Information structures, types, 46
Information superiority, 97
Information triage, 4, 191
Initial response, 138
Insight, 15f
International Civil Aviation Organization, 11
Intentions, shared, 195
Interaction strategies, 46–47
Inter-agency communications, 140t
Internal representations, 33–34, 40, 51
Interoperability
 analysing network structures and, 184–189

challenges of, 167–179
    case study: Boston Marathon Bombing (2013), 176–178, 177f
    case study: London Bombing (2005), 173–176, 174f, 175f
    conclusions, 178–179
    introduction, 167–168
    definition, 168
    emergency services, 225
    Interoperability Continuum, 168–173
        governance, 169f
        overview, 168–169, 169f
        standard operating procedure (SOP), 169f, 170–172
        technology, 169f, 169–170
        training, 169f, 172–173
        usage, 169f, 173
Interoperability Continuum, 168–173
    governance, 169f
    overview, 168–169, 169f
    standard operating procedure (SOP), 169f, 170–172
    technology, 169f, 169–170
    training, 169f, 172–173
    usage, 169f, 173
Inter-service collaboration, 214
Iran Air Flight 655, 90
Issue focusing, 51

J

Jefferson, T., 190
Jenkins, D. P., 36
Jentsch, F. G., 98
JEWG, *see* Joint Engagement Working Group (JEWG)
Johannesen, L., 63, 64
John F. Kennedy Library, 177
Johnston, R., 70, 71

Joint Engagement Working Group (JEWG), 18
Jones, W. M., 66
Just-in-time mental models, 16

K

Kahneman, D., 33
Khalilbeigi, M., 43
King's cross, 175f, 242, 243
Kirsh, D., 40, 42, 43
Klein, Gary, 15, 26, 27, 28, 45, 63, 96, 97, 98
Knowledge, 23
    by acquaintance, 23
    actionable, 202
    by description, 23
    as modified and produced, 69
    mutual, 64

L

Landgren, Jonas, 22, 59, 61, 202
Language, 58, 123
Law enforcement co-ordination centres, 176
Leedom, Dennis K., 31
Liaison officers (LO), 154–155, 164, 214
Linguistic evidence, 10, 11
LO, *see* liaison officers (LO)
Local authority, 76
Local controllers, 108, 110, 133
Local radio talk group, 110
Logical deduction, 15
London Ambulance Service, 175f, 176, 185f, 187–188, 245
London Bombings (2005), 36, 173–176, 174f, 175f, 182, 205, 241–244
London Fire Brigade, 175f, 187, 188, 245
London Underground, 64, 185f, 245

Loosely coupled work systems, 91, 91t
Low-tech artefacts, 42, 43
Luff, P., 64

M

McMaster, Richard, xix, 17
Macrocognition, 15–16, 15f, 92t
Macro-organisation, 194
Maglio, P., 40
Major incidents
   challenges, 139–141
   definition, 138–139
   issues, 139
   phases, 138
   reflection, 223–225
   routine emergencies versus, 139
   responding, 137–150
      C2 Structures, 143–145
      introduction, 137–141
      tracking situations, 145–149
      Walham Floods (2007), 141–143
Manning, Peter K., 3, 71
Maps, shared, 196
Media, external, 57
Medical emergencies, 102
Memory, communal, 47
Menezes, Jean Charles de, 36
Mental simulation, 24, 25
Metropolitan Police Service control, 185f, 187, 245
Military, 151, 152, 154–155, 156t
Military assistance, 154n
Misovich, S. J., 23
Modality, 18t
Models, systems dynamics, 71
Multi-agency emergency response, 140–141, 140t
Multi-agency major incident doctrine, 162
Mutual aid scheme, 143

N

9/11 attack, 95
999 calls
   abbreviations and acronyms, 106
   callers, 104–105
   call handlers, 104–106, 105f
   dispatcher, 104–105
   FCC, 79
   open questions, 105–106
   process flow, 107
   routine incident response, 122t
Narratives, 28, 117f, 122f, 128f, 129, 220
National decision model (NDM), 81f, 81
National Grid, 152, 153
National Health Service, 76
National Police Improvement Agency (NPIA), 194
NATO, *see* North Atlantic Treaty Organization (NATO)
Naturalistic decision-making (NDM), 21
Navigation, 44–45, 45f
NDM (national decision model), *see* National decision model (NDM)
NDM (naturalistic decision-making), *see* Naturalistic decision-making (NDM)
Negotiation, 200
Network diagram, social, 186
Network structures analysation, 184–189
New York City, 95
New York Times, 1
Non-urgent communications, 110
North Atlantic Treaty Organization (NATO), 78
Norway, Oslo bombing, 1–2
Notebook, 129

INDEX

Notepad, 121, 122t, 122f, 123, 128t, 129
Notepaper, 123, 124f
Novices (sense makers), 7, 16; *see also* Individual sensemaking
NPIA, *see* National Police Improvement Agency (NPIA)

O

Objects, multiple manipulation, 47
Observe, orient, decide, act (OODA) loop, 82, 83f, 84
OCU, *see* Operational command unit (OCU)
Offenders, 108
Officer emergency assistance, 110
Officers, *see* Personnel; Police officers
OODA loop, *see* Observe, orient, decide, act (OODA) loop
Open-ended questions, 105–106, 122f
Operational command unit (OCU), 79, 80f, 103f, 103, 122t
Operation Telic (2003), 90
Organisational structures, emergency response, 162–163
    analysing network structures and interoperability, 184–189
    challenge of sharing information, 190–192
    sensemaking as a social process, 182–184
Omerod, T., 4–5
Oslo bombing, 1–2
Otto's notebook, 52–53

P

Perceptual evidence, 10, 10–11
Perry, M., 43, 44, 133

Personal protective equipment (PPE), 153
Personnel, 103f, 144f, 182, 202
Perspectives, shared, 195
Pirolli, P., 49
Planning and adaptation, 93–96
Plans, definition, 30–31
PNC, *see* Police National Computer (PNC)
Pocket notebook
    as artefacts, 122t
    formal purpose, 128t
    IMS versus, 120
    police incident response sensemaking, 128t
    sensemaking role, 128t
    function, 114–115
    sensemaking in suspect's identity, 116–118, 117f
    significance, 116
Police National Computer (PNC), 106, 108, 118, 123, 124f, 126
Police officers
    frame-defined data collection, 110, 111, 112
    improvised artefacts, 131
    as initial responder, 75
    resource allocation to incidents, 125, 126–127
    responding, 122t
    shifting of roles, 202
Police response activity, 4, 36, 62, 123
Posey, M., 98
Potter, S., 94
Practice, competitive, 140t
Presentation, 50, 122f
Prior experience representation, 23–27
Problem detection, 15f, 28, 96–98
Problem space, 214

Pseudorationality, 3
Pseudo silver, 144, 144f, 145
Published entries, 132

R

Radio, *see* Digital radio
Radio systems, 151
Radio talk groups, 48, 118–119, 133, 213
RAF, *see* Royal Air Force (RAF)
Rahman, A., 140
Ramduny-Ellis, D., 47
Rationality, 3
Recognition-primed decision (RPD), 24–27
    model, 24–25, 25f
    research, 26
    sensemaking and decision making, 26–27
    significance, 25–26
Recovery, 138
Reddy, M., 42, 43, 129
Redundancy, 58
Representation construction, 49–52, 50f
Representations
    call handler's sensemaking process, 122f
    common, 58, 59
    cyclical view, 46–47
    external, 33–34, 40, 42–43, 50, 51, 127
    internal, 33–34, 40, 51
    shared, 195
    transformation of, 41, 58
Rescue service commanders, 73n
Resource allocation
    controllers, 123–126
    police officers, 125, 126–127
Resource allocators, 130
Resource for action, 18t

Resources
    allocation, 123–127
    allocators, 130
    deployment and mobilisation, 78, 79
    for action, 18t
Responding officers, 122t
Responding units, 130
Response co-ordination, immediate, 110
Restoration, 138
Retail thefts, 102
Richard, Martin, 176
Right of interpretation of the situation, 59
Risk management, 15f
Road traffic incidents, 102
Rochlin, G. I., 90
Routine emergencies, 208–210, 221n
Routine incidents
    artefacts in, 164
    distributed cognition
        allocating resources to incidents, 123–127
        controllers, 123–126
        police officers, 125, 126–127
        call process, 121, 122f, 123
        conclusions, 134–135
        improvised artefacts, 131–132
        making sense through artefacts, 129–131
        making sense through collaboration, 132–133
        making sense with artefacts, 127–129
        routine incident response, 121, 122t
    management, 101–120
        call process, 104–106, 107f
        challenges, 102
        closing the incident, 120
        examples of routine incidents, 101–102

incident response C2, 102–103
introduction, 101–104
officer attending, 114–119
supporting responding units, 106, 108–114
frame-defined data collection, 110–112
response, 121, 122t
collaborative sensemaking and, 135
sensemaking, 129–131
Royal Air Force (RAF), 142
Royal National Lifeboat Institution (RNLI), 147–149, 152, 154
RPD, *see* Recognition-primed decision (RPD)
Russell Square, 175f, 175, 176, 243

S

7 July Review Committee, 175f, 176, 178, 184, 186
SA, *see* Situation assessment (SA)
SCG, *see* Strategic co-ordination group (SCG)
Schema
 definition, 23–24
 function, 24
 distributed SA and, 65–67
 data-frame model and, 28
 frames and, 28n
 representation construction, 49–52, 50f
Schemata, 23–24
Scripts, 28n
Seeing the gap act, 22
Sense, 6–8, 211
Sense makers, 7, 63
Sensemaking
 artefacts in police incident response, 128t
 breakdown during major incident, 213–214
 call handler's, 122f
 challenges, 1–8
  forms, 6–8
  occurrence of sensemaking, 5–6
  Oslo bombing (2011), 1–2
  recognition of information, 2–5
 characteristics, 2–5, 15–16, 15f
 command and control (C2), 89–100
  collaborative networks, 91–93
  conclusions, 99–100
  co-ordination, 98–99
  introduction, 89–91
  planning and adaptation, 93–96
  as prerequisite, 79
  problem detection, 96–98
 common ground and, 7, 11–12
 common ground in conversations, 8–12
  overview of common ground, 9–11
  phone call example, 8–9
  sensemaking and common ground, 11–12
 as continuous process, 61–62
 decision making and, 26–27
 distribution cognition, target review process, 18, 19f, 20
 as distributed cognition, 219–221
 distributed cognition
  features of activities in task, 18t
  graphical representation, 17–20
  military exercise, 17–18
  overview, 16–17
 enactment process, 60–61
 individual commander and, 27
 macrocognition and, 15–16, 15f
 occurrence of, 5–6

INDEX

organisational structure, emergency
response, 181–192
  analysing network structures
and interoperability, 184–189
  challenge of sharing
information, 190–192
  conclusions, 189–190
  introduction, 181–182
  sensemaking as a social
process, 182–184
plausible rather being/than true,
62–64, 68
police incident response, 128t
as representation construction,
49–52, 50f
retrospective process, 60
routine emergencies, 212–213
routine incident response,
129–131
sense and, 6–8
situations, 5–6
as social, 61
as social process, 182–184
suspect's identity, 116–118, 117f
as system activity, 70–71, 70f
types, *see* Artefact-driven
sensemaking; Collaborative
sensemaking; Individual
sensemaking
September 11 attack, 95
Serfaty, D., 94, 95, 98
Shoebox, 50f, 50
Silver (tactical) command; *see also*
Walham Floods (2007)
  Boston Marathon Bombing
(2013), 176
  Force Communications Centre
(FCC), 79
  function, 75, 76, 78f
  hierarchical structure, 74f, 78f
  national decision model (NDM),
81f, 81
  pseudo command, 144, 144f, 145

Situation, 5–6, 212
Situation assessment, 24, 27, 64
Situation awareness (SA)
  definition, 64–65
  distributed, 65–69
    role of artefacts, 69
    SA approaches, 65–66
  distributed SA
    distributed concept, 68–69, 68f
    shared concept, 66–68, 67f
  personnel from difference
agencies, 182
  pragmatic aspects, 69
  semantic aspects, 69
  team, 66
Situation space–decision space gap,
206
Socially distributed cognition, 133
Socially shared cognition, 57–59
Social network analysis centrality
scores, 245–246
Social network diagram, 184, 185f
Social organization, 58
Social processes and collaborative
sensemaking, 163–165
SOP, *see* Standard operating
procedures (SOP)
South Asian Tsunami (2004), 140
Speed bugs, 44, 45f
Standard operating procedures (SOP),
86, 169f, 170–172, 212
Stanton, N. A., 68, 84, 102
Stocks, 71
Stone, N. J., 98
Storage, 18t
Strands of expert, 22f
Strategic coordinating group (SCG),
77, 143
Street robberies, 102
Structured articulation, 42
Subtle cues, 96
Suchman, L. A., 46
Suicide, 102

Systems dynamics model, 70, 70f, 71
Systems-level cognition, 41

T

Talk group communications, 112–113, 118–119, 213
Target sheets, 18
Task–artefact cycle, 12f
Tasks, distribution of, 49
Taskwork knowledge, 68–69
Tavistock Square, 175f, 186, 244
Taylor, J. R., 23
Team co-ordination, 69
Team mental models (TMM), 68–69
Teamwork knowledge, 69
Technology, 4, 169f, 169–170
Tightly coupled work systems, 91, 91t
TMM, see Team mental models (TMM)
Traditional hierarchical command and control (C2), 92t, 92, 205
Traffic controllers, 103f, 110, 122t
Training, 169f, 172–173
Transactive memory, 14, 68
Transformations
  information, 18
  representations, 41, 58
Transmission, 18t, 110
Trust, 140, 140t
Tversky, A., 33

U

UCC, see unified command centre (UCC)
United Kingdom (UK)
  command and control, 73–87
  concept, 78–86, 80f, 81f
  emergency service operations, 74–78, 74f
  future, 86–87
  introduction, 73–74
Crime Survey for England and Wales, 34
Emergency Services, 168, 181
England, 75, 142
London Ambulance Service, 175f, 176, 185f, 187–188
London Bombings (2005), 36, 173–176, 174f, 175f, 182, 205, 241–244
London Fire Brigade, 175f, 187, 188
London Underground, 64, 185f
Police National Computer (PNC), 106
Wales, 34, 75, 142
Warwickshire Police, xix, 73n, 80f, 101, 110, 124, 124f, 125, 126
West Midlands Ambulance Services (WMAS), 79
West Midlands Fire Service (WMFS), 79
West Midlands Police (WMP), xix, 73n, 79, 80f, 101, 125, 108, 108f
Umapathy, K., 64
Uncertainty management, 15f
Unexpectedness, 5
Unified command centre (UCC), 176–177
Unmanageableness, 5
Unprecedentedness, 5
Unstructured articulation, 42
Urgent welfare concerns, 102
Usage, equipment, 169f, 173
US Department for Homeland Security, 168, 190
USS Vincennes, 90
Utøya massacre, 1

V

Van Fenema, P. C., 99
Vehicle crime, 102
Vocabulary, 77, 191

W

Wales, 34, 75, 142
Walham electricity substation, 141
Walham Floods (2007), 141–143, 142f, 182–183
Walker, G. H., 67
Warwickshire Police, xix, 73n, 80f, 101, 110, 124, 124f, 125, 126
Weick, Karl Edward, 5, 7, 57, 63, 68, 71, 97, 99, 165
Wenger, Etienne, 14, 68
West Midlands Ambulance Services (WMAS), 79
West Midlands Fire Service (WMFS), 79
West Midlands Police (WMP), xix, 73n, 79, 80f, 101, 125, 108, 108f
Westrum, R., 97
Whalen, M. R., 8
WMAS, *see* West Midlands Ambulance Services (WMAS)
WMFS, *see* West Midlands Fire Service (WMFS)
Wolbers, J., 200, 202
Wong, B., 65
World Trade Center, 95
Wright, P. C., 42, 46
Wu, A., 195, 203

Z

Zimmerman, D. H., 8